M000289744

This book passed through
LFL# 54236
Vacaville, CA

IMAGES
of America

SAN FRANCISCO'S
PRESIDIO

The Presidio's entire 1,480 acres can be seen in this February 29, 1932 aerial photograph taken by Army 15th Photo Section, stationed at Crissy Field. When approaching the Presidio by air from the east, the Palace of Fine Arts, Crissy Field, Letterman Hospital, Fort Winfield Scott, and golf links are recognizable landmarks. (Courtesy of the National Park Service, Golden Gate National Recreation Area, Park Archives & Records Center GOGA-4013.)

IMAGES
of America

SAN FRANCISCO'S
PRESIDIO

Robert W. Bowen

ARCADIA
PUBLISHING

Copyright © 2005 by Robert W. Bowen
ISBN 978-1-5316-1607-6

Published by Arcadia Publishing
Charleston, South Carolina

Library of Congress Catalog Card Number: 2005922904

For all general information contact Arcadia Publishing at:
Telephone 843-853-2070
Fax 843-853-0044
E-mail sales@arcadiapublishing.com
For customer service and orders:
Toll-Free 1-888-313-2665

Visit us on the Internet at www.arcadiapublishing.com

This book is dedicated to the memory of my grandparents Frank and Tillie Bowen;
to my parents, Robert and Lucy Bowen; and my sister Linda Bowen.

"We went through the Presidio this afternoon. It extends along the bay for two or three miles.
It is where the soldiers are stationed." This postcard is postmarked May 23, 1909.

CONTENTS

ACKNOWLEDGMENTS

Writing a book is a challenge, and would be an impossible task without professional assistance, support, and contributions from a multitude of special people.

First and foremost, I would especially like to thank my wife, Brenda, whose love, encouragement, patience, and willingness to sacrifice time and space more than anything else made this book possible.

I would particularly like to thank National Park Service ranger Will Elder for his considerable assistance. Ranger Will has been my mentor and an invaluable resource to me in all things related to the Presidio as a national park. Will has also provided much technical advice and support whenever help was needed. Serving as Will's intern in 2004 was an incredible opportunity and learning experience.

I would especially like to thank my other mentor with the National Park Service, Ranger Marcus Combs. Marcus's dedication and enthusiasm for interpreting the story of the Presidio is boundless.

Special thanks to John Poultney, my editor at Arcadia Publishing, for seeing the need for this book and his forbearance in guiding the project to completion.

I would like to express my gratitude to Susan Ewing Haley of the National Park Service, Golden Gate Recreation Area, Park Archives and Records Center for her practical advice and speedy assistance. I would also like to thank Richard Marino, photograph curator, and Susan Goldstein, city archivist, of the San Francisco Public Library History Center for their assistance. All items not credited are from the author's personal collection of postcards, photographs, and Presidio memorabilia.

My thanks to Randolph Delahanty, historian for the Presidio Trust, and Stephen Haller and John Martini, historians for the National Park Service. Books, lectures, and presentations by these knowledgeable historians provided me with an opportunity to take notes and acquire particular information and insight regarding the story of the Presidio. The late Erwin N. Thompson's two-volume history of the Presidio, *Defender of the Gate*, is the ultimate resource and account about the Presidio, and was a reference that I relied upon greatly.

I would also like to express my thanks and appreciation to the following individuals who contributed some information or assistance toward the completion of this book: Woodrow Schenebeck, a hospital train veteran; Ed Herny of the San Francisco Bay Area Post Card Club; Paul Rosenberg; Luca DiDonna; Therese Poletti of the *San Jose Mercury News*; Jan Chaffee of the Interfaith Center at the Presidio; Don Gray of the Crissy Field Aviation Museum Association; Johann Kingsfield and all my fellow docents at the Presidio; Tyler Gee of Fireside Camera; Gama Photography; Chris Bowen; park rangers Dan Ng, Rik Penn, and James Osborne; and retired ranger Janice Cooper.

I would also like to express my gratitude to the all the rangers, park volunteers, the Presidio Conservancy, and the members of the Fort Point and Presidio Historical Association for their interest in keeping the story of the Presidio alive.

Finally, I would like to express my thanks and gratitude for the contributions and service of the thousands upon thousands of soldiers who passed through the gates of the Presidio of San Francisco.

INTRODUCTION

In the spring of 1776, a Mexican frontier soldier named Juan Bautista de Anza led an expedition to the isolated, sandy bluffs overlooking San Francisco Bay. Seeing that the site with its sweeping view of the bay would be excellent for protecting their territory of New Spain, Captain Anza and his party erected a wooden cross, marking their claim for the Spanish empire.

On September 17, 1776, this sparse land was formally established by Anza's lieutenant, Jose Joaquin Moraga, as the third of four outposts in Upper California. Moraga arrived with a band of nearly 200 soldiers and civilian settlers. When they arrived the land was a mass of rocky, sandy hills, sparsely covered by shrubs, grasses, laurel trees, and swampy marshes. The Ohlone who occupied the peninsula for hundreds of years had used the desolate spot to hunt small game, fish, and gather shellfish.

The soldiers for New Spain constructed a presidio, or fort: a temporary mud and thatched quadrangle approximately 200 yards square surrounded by a palisade wall. Eighteen years later, under the threat of British territorial claims on the West Coast, the Spanish built another fortress on the steep bluffs, the Castillo de San Joaquin, using cannons to guard the strait into San Francisco Bay.

As the Spanish became more involved in European wars and wars of independence by their American colonies, the adobe fort suffered from neglect. When the fort was battered by storms and earthquakes, little was done to repair the damage. Mexico gained independence from Spain and control of the Presidio in 1822. The Mexican government continued the policy of neglect. More concerned with overland American expansion into California, they eventually transferred their troops north to Sonoma.

The American story of the Presidio began in 1846, when the United States Congress declared war on Mexico. During the Bear Flag Revolt, explorer John C. Fremont was determined to disable the guns of the Castillo overlooking the bay. On July 1, 1846, Fremont and his men sailed across the bay from Sausalito to the abandoned Presidio and drove metal files into touchholes of the rusty cannons, rendering them totally useless.

The formal occupation of the Presidio by American troops took place in March 1847 with the arrival of two companies of Stevenson's New York Volunteers, commanded by Maj. James Hardie. The first regular Army troops, led by Capt. Erasmus D. Keyes, arrived in May 1849. Within a few weeks, most of the 86 men in the detachment deserted to the gold fields near Sacramento, leaving Keyes, who would remain in charge of the Presidio off and on for the next 10 years, with nary a soul under his direct command. This was the beginning of the Presidio's 147 years of auspicious history as one of the foremost military posts in the United States.

In the 1850s, the Army began a major rebuilding of the old, dilapidated adobe sites of the Presidio and its surroundings. The building construction continued throughout the 19th century. Many of those buildings survive today, providing a unique San Francisco neighborhood of historic architecture. In the 1880s, Army engineer Maj. William A. Jones recommended to Maj. Gen. Irvin McDowell that large areas of the barren post be planted with pine and eucalyptus tress. Those trees were planted and ultimately grew to give the Presidio its lush, park-like appearance we enjoy today.

The history of the Presidio reflects the events that occurred in the nation, the state, and the city. Soldiers from the Presidio played a role in the wars against the Native American Modoc and Apache, and in the wars of the 20th century from the Spanish-American to the Vietnam War. Modern developments in military science are found in the coastal artillery batteries with their huge guns and early air power at Crissy Field. The Presidio provided a haven for earthquake refugees in 1906, the excitement of a world's fair in 1915, and access to the Golden Gate Bridge beginning in 1937. For over 90 years Letterman General Hospital provided excellent medical care for thousands of military personnel and their dependents. In 1960, the Presidio of San Francisco was designated as a National Historic Landmark.

The Golden Gate National Recreation Area was established in 1972, with Fort Point, the batteries, and the shoreline placed under the auspices of the National Park Service. In 1994, the Presidio was turned over to the Park Service with a caveat by Congress that the new park would be financially self-sufficient by 2013. The Presidio Trust was created in 1996 to manage the funding, most of which has come from the leasing of both historic and new buildings. Today, the Presidio is one of the largest historic preservation projects in the United States. The Presidio Trust has been responsible for rehabilitating more historical buildings than any other site in the country. The new Letterman Digital Arts Center is on the site of the Letterman Hospital and is leased to filmmaker George Lucas of *Star Wars* fame.

During its 218-year history as a military post, no shot was ever fired in anger. The Presidio was a popular assignment for officers, noncommissioned officers, soldiers, and their families. Today it is a unique national park and a new vibrant residential and commercial neighborhood in a very special city—San Francisco, California.

One

19TH-CENTURY
GARRISON

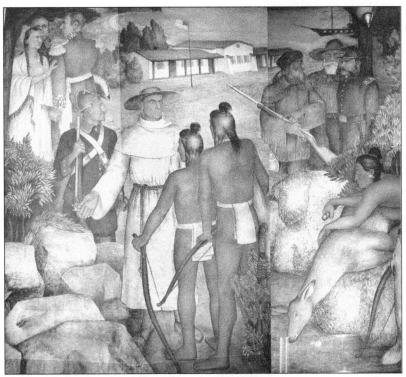

On the east wall of the Presidio's post chapel, overlooking San Francisco Bay, there is a large mural. Painted in 1935 by noted artist Victor Arnautoff (1896–1979) as a California Relief Administration project and entitled *The Peacetime Activities of the Army*, the mural illustrates the early history of the Presidio. It depicts Ohlone hunters who originally used the land. A padre and a soldier present the Spanish cross and sword. Tragic romance is depicted by 15-year-old Concepcion Arguello holding the hand of the dashing Nikolai Rezanov, chamberlain to the Russian czar. A scout, miner, and blue-coated Army officer represent the arrival of the Americans at the Presidio.

John Charles Fremont (1813–1890), a pathfinder and officer in the United States Topographical Corps, led a group of "Bear Flag insurrectionists" on July 1, 1846, and symbolically spiked the 10 guns of the Castillo de San Joaquin. Fremont coined the name Chrysopylae, or "Golden Gate," for the entrance to the bay. He would use his notoriety to be elected to the United States Senate and to win the nomination as Republican candidate for president in 1856.

This is one of the bronze cannons from the Castillo de San Joaquin that John C. Fremont believed could be used to harass American ships. The guns were cast in Lima, Peru, in 1673 and brought to the Presidio in the 1790s. Two cannons, named Poder and San Pedro, now sit as decorations in front of the Presidio Officers' Club with their touchholes permanently sealed and forever silent.

The Presidio of San Francisco.

The flag of the United States flies over the Presidio in this 1850s woodcut. In the early years of American occupation, the Army claimed that the original boundaries of the Presidio extended from the western edge of the peninsula east to the area of today's Fisherman's Wharf. The present boundaries of the Presidio were approved by Pres. Millard Fillmore in 1851. (Courtesy of the National Archives III-SC91387.)

This is an image of Winfield Scott (1786–1866), general-in-chief of the United States Army when American troops first occupied the Presidio. Scott was a mentor to Capt. Erasmus D. Keyes, the first regular Army officer to command the post. Scott visited San Francisco in 1859 where he was greeted with fanfare and salutes by the 3rd Artillery. The sub post at the northwest corner of the Presidio would eventually be named in Scott's honor.

11

Fort Point is viewed from the ocean side in this 1870s cabinet photograph by Thomas Houseworth. The military always recognized the strategic importance of San Francisco Bay. In 1853, Army engineers chose to build a brick fortress and seawall near the site of the old Spanish Castillo de San Joaquin. The headland, Punta del Cantil Blanco, was reduced from an elevation of 97 feet to 16 feet. Original plans called for a total of 142 weapons to be located at the fort. Completed in 1861, Fort Point was heavily armed but within four years, technical advances in artillery ordnance would render the brick and mortar fort obsolete. National Park Service historian Edwin C. Bearss, in the Fort Point *Historic Structure Report*, noted that the fort's 1867 weapons inventory consisted of six 24-pounders, eleven 32-pounders, thirty-eight 42-pounders, eight 8-inch columbiads, and two 10-inch columbiads. The guns were seldom used except to fire salutes on national holidays. A lighthouse with a Fresnel lens was set on the barbette tier of the fort in 1864 to provide a beacon for ships passing through the Golden Gate.

Viewed from the sea, Fort Point presented a formidable image as a defender of the San Francisco Bay. The fortress continued to be armed until the late 1890s when it was finally replaced by new reinforced concrete coastal batteries. When the Golden Gate Bridge was constructed, chief engineer Joseph B. Strauss, an admirer of the beauty of Fort Point, had a special bridge arch constructed to prevent the fort from being demolished.

The post General Hospital was built in 1864, replacing a smaller hospital building. Four years later, Eadweard Muybridge photographed the hospital with two recovering soldiers standing near the parade ground with crutches and a cane. Named in honor of Brig. Gen. George Wright, the hospital had beds for 50 patients and 10 rooms, including a kitchen, dining room, morgue, and prisoners' ward. In 1878, the hospital was reoriented, facing toward the east along with the homes on officers' row. (Courtesy of the San Francisco History Center, San Francisco Public Library.)

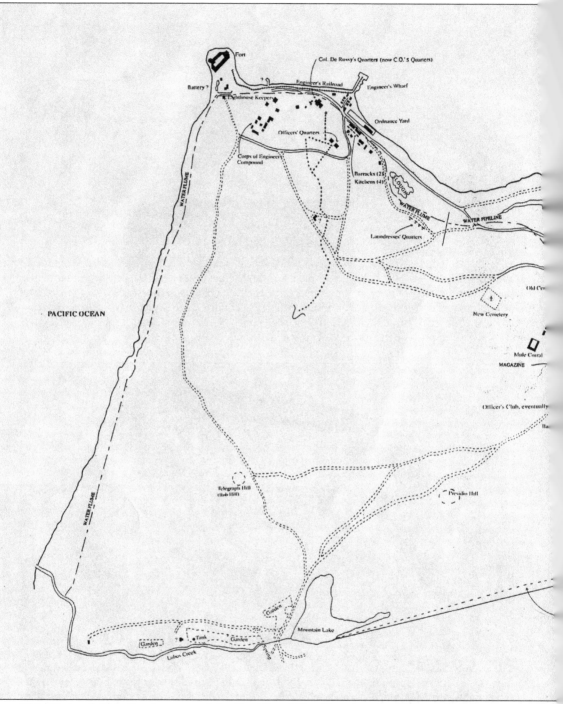

This National Park Service map of the Presidio of San Francisco in 1870 was based on an original map by Lt. George M. Wheeler of the Corps of Engineers. Wheeler was responsible for a survey of the reservation to help resolve the issue of land disputed by both the Army and the city of San Francisco. The triangle of land relinquished by the Army is on the right along today's Lyon Street. The Army retained the bottom claim along what is now West Pacific

PRESIDIO OF SAN FRANCISCO
1870

reference north

SCALE IN FEET
0   250   500   1000   2000

SAN FRANCISCO BAY

FINAL EASTERN BOUNDARY

Presidio Wharf

Private Wharfs

Harbor View Resort

Private Development

SLOUGH

SLOUGH

WATER PIPELINE

Stables

Future Drill Field

TRIANGLE OF LAND EVENTUALLY SURRENDERED BY ARMY

Barracks

Hospital

Officer's Quarters

MAIN ROAD TO SAN FRANCISCO

Presidio House

HDQ

Chapel

Rancho Ojo de Agua de Figueroa Claim

Pond ?   Hill

Queen Springs

...CLAIMED BY CITIZENS BUT RETAINED BY ARMY

LEGEND

~~~~~   Road
········   Trail
•••••   Water Pipe Line
~~~~~   Stream
~~~~~   Slough
+++++   Railroad
— — —   Water Flume

NOTE:
1. From a map by Lt. George M. Wheeler.

Avenue. The sloughs are shown in the northeastern area of the post, near where the man-made tidal marsh is now located. Water pipelines and flumes, roads, and more then 55 buildings comprise improvements the Army had made on the post since 1849. The new cemetery would become the San Francisco National Cemetery. (Courtesy of the National Park Service, Golden Gate National Recreation Area Park Archives and Record Center.)

Citizens of San Francisco on July 3, 1876, were invited onto the grounds of the Presidio to join the military in a celebration of the nation's centennial. In this view, civilian spectators on a treeless plateau observe a "sham battle" reenacted by California National Guardsmen. Troops playing the role of "enemy" proceeded from the Presidio House, the two-story white building in the middle of the photograph. The road house stood just beyond today's Lombard Street Gate within the disputed land triangle. The neighboring area, consisting of several dairy farms and Chinese vegetable gardens, was known as Harbor View for a local resort. There is no record the cows were disturbed during the mock battle as infantry and cavalry noisily dashed back and forth, creating a residue of smoke and dust. The following day, D Troop of the 1st Cavalry and four batteries of the 4th Artillery joined the National Guardsmen as they marched through the streets of San Francisco in a great parade celebrating the nation's 100th Fourth of July. (Courtesy of the San Francisco History Center, San Francisco Public Library.)

This 1890s view of the main post with its well-groomed military look shows the physical changes to the landscape and the major building constructions of the previous decades. The grassy parade ground is crisscrossed by pathways. The two-story commanding officer's home is on the left, opposite the flag pole. Facing the parade ground are a row of six single-story barracks and two barracks that have been altered to provide an additional second floor. Two two-story wood frame artillery barracks face the parade ground from the north. The houses on officers' row have been reoriented to face east and a fence protects each home's lawn and flower garden. The large white building, also known as "the corral," is the Bachelor Officers' Quarters. The next building is the post chapel and nearby is the Officers' Open Mess, built from the old Spanish adobe commandant's house. The most significant changes to the landscape are the rows of pines trees planted in the foreground that are the beginnings of the Presidio forest. During the celebration of the first Arbor Day in 1886, school children planted 3,000 tree slips donated by local philanthropist Adolph Sutro. (Courtesy of the National Park Service, Golden Gate National Recreation Area Park Archives and Record Center.)

The post commander or senior ranking officer occupied this prominent residence. The large Victorian contained all the amenities for fine 19th-century living and entertaining. This 1887 view shows the home surrounded by a white picket fence. In the foreground is a mounting block that provided access to and from carriages. In 1915, fire swept through the home when a hot cinder fell onto the highly polished hardwood floor, killing Gen. John J. Pershing's wife, Frances, and their three daughters. (Courtesy of the National Park Service, Golden Gate National Recreation Area Park Archives and Record Center.)

The 12 homes on officers' row were built in 1862 and occupied by senior officers and their families. The homes contained a parlor, sitting room, three bedrooms, dining room, kitchen, buttery, washroom, bathroom, and a water closet. Originally facing the parade ground in 1878, the houses were remodeled to face east on what is now Funston Avenue. In 1947, the buildings were converted to duplexes.

Eight Civil War–era enlisted barracks were erected on the west side of the parade ground to provide quarters for companies of infantry and later cavalry troops. In 1878, the first two buildings became headquarters for the Military Division of the Pacific, and later, all the buildings were used as offices. In 1885, the two northernmost buildings were altered to add an additional story. Electricity was installed in 1912.

This picture of the parade ground and red brick barracks, by post photographer James D. Givens, provides a view of the Presidio at the turn of the 19th century. Research by historian Erwin N. Thompson indicated that post commander Col. William M. Graham felt that because of the size and importance of the Presidio and its proximity to a major city, new barracks should be constructed of modern materials and be of ample size. The barracks were completed by 1897. The white building behind the barracks was the former home and store of the post sutler. The athletic grounds to the south gave troops a place to play baseball, football, and participate in equestrian sports.

The end of an era is evident in this 1894 photograph by amateur photographer I.S. Foorman of the California Camera Club. A solitary soldier stands by a Civil War–era 24-pounder field howitzer near Fort Point—a classic example of mid-19th-century coastal fortification already obsolete by the time the fortress was completed. Seventy-five years later, the Fort Point and Army Museum Association, a nonprofit organization now known as the Fort Point and Presidio Historical Association, lobbied Congress to save the decaying old fort. Identical bills were introduced by San Francisco congressmen William Mailliard and Philip Burton, and by United States senators George Murphy and Alan Cranston. On October 16, 1970, Pres. Richard Nixon signed Public Law 91-457, making Fort Point a National Historic Site that would be restored and managed by the National Park Service.

Two

SPANISH-AMERICAN AND PHILIPPINE WARS

The United States Army was ill prepared for war when the Battleship USS *Maine* blew up in Havana harbor on February 15, 1898. Two and a half months later Congress declared war on the kingdom of Spain and Pres. William McKinley called for 125,000 volunteers. It took the United States just four months to defeat the Spaniards, but a direct result of that victory was the Philippine War, which lasted until 1902. The Presidio of San Francisco played an important role as the United States extended itself militarily west across the Pacific. The Presidio became extremely active as thousands of soldiers converged in preparation for embarkation to Hawaii, Guam, the Philippines, and China. This photo taken during the summer of 1898 shows the camps of the 51st Iowa and 1st New York Volunteer Infantry.

The first volunteer troops arrived at the Presidio in June 1898. Camp Miller was located in the Presidio near the eastern boundary of the post. Camp Merritt was located outside the Presidio, bounded by First Avenue (Arguello), Sixth Avenue, Fulton Street, and Balboa Street. Camp Merriam occupied the low ground east of officers' row. Soldiers were quartered seven to eight men in a tent and spent hours each day in drill.

This cavalry camp in Tennessee Hollow was part of Camp Merriam. The 1st Troop, Utah Volunteer Cavalry, occupied the foreground. The tents of the Nevada Volunteer Cavalry are located behind the 1st Utah. Before being mustered out in 1898, the Utah volunteers were sent to superintend Yosemite and Sequoia National Parks. (Courtesy of the National Park Service, Golden Gate National Recreation Area Park Archives and Record Center.)

A souvenir booklet of Army life featured this photograph of soldiers eating at their outdoor mess facility. The Presidio soldiers were well fed, with rations of beef, bacon, fresh fish, bread, coffee, and canned as well as fresh vegetables. The Army also had travel rations that were often used on quick marches: one pound of hard bread, three-quarters of a pound canned of beef, one-third of a pound of canned baked beans or tomatoes, coffee, and sugar.

Another souvenir view shows six soldiers drinking in a secluded spot near their camp. Drinking was prohibited for enlisted men and the canteen on post was abolished, but nearby grocers and saloonkeepers were happy to accommodate thirsty soldiers. According to historian Erwin Thompson, the troops' favorite beverage was San Francisco's Buffalo beer, especially a brew named Columbacher. The major problems caused by alcohol were public drunkenness, fighting, and rowdy behavior, but there is little evidence that off-duty soldiers were a serious public nuisance, and most men conducted themselves appropriately.

One of the most popular men in camp was the battalion mail clerk. Writing letters and receiving mail from home played an important role in keeping up soldiers' morale, as seen in this souvenir view.

Patriotic envelopes and stationery were distributed to the troops. This letter from Pvt. Ray Owen of Company M, 20th Kansas Volunteers, was mailed from the Presidio on October 4, 1898, to his brother Guy Owen in Kansas. Private Owen described his life in camp:

> The Presidio is a fine place. We gave an exhibition downtown, it commenced to rain as we left camp. We had to walk two miles to take the streetcar. The entertainment was for the benefit of a widow who lost her husband at Manila. They took in between $3,000 and $4,000 for her. . . . I have got to smoking a pipe; there are 14 of us in a tent and we all smoke. I tell you we fill the old tent full of smoke. . . . There are about 8,000 troops in the Presidio, we have some good sham battles and some target practice. I tell you the old rifle kicks like the devil. Some of the boys had the skin kicked off their shoulders.

Local entrepreneur Edward H. Mitchell had thousands of picture postcards printed showing Army life in San Francisco. The Private Mailing Card Act of Congress passed on May 19, 1898, coincided with the opening shots of the Spanish-American War. It cost a soldier only 1¢ to mail a postcard greeting from the Presidio.

This early postcard was entitled, "Off to the Wars! U.S. Troops Leaving Jackson and Lyon Streets, Entrance of Presidio, San Francisco." In their haste to publish this postcard, local San Francisco printers Britton & Rey misidentified the Lombard Street Gate. The 51st Iowa Volunteers, bound for Manila, marched out of the Presidio on their way to board SS *Pennsylvania* on November 3, 1898.

The tents shown in this 1902 postcard are in the east cantonment area of the Presidio. There were camps for troops en route to the Philippines and another for those returning. Two additional camps were set up for casuals and recruits undergoing training and awaiting transportation overseas. A detention camp was established for recruits arriving with infectious diseases.

The tents in the foreground show a model camp for officers either en route to or returning from the Philippines. The parade grounds with the officers open mess, offices, and barracks are clearly visible. In 1906, postcard publishers misidentified the tents in this photograph as an earthquake refugee camp.

This souvenir view of two soldiers returning from the Philippines greeting their sweethearts at the Presidio was part of a photographic series of Army life.

The original caption on this turn-of-the-century photo best expresses how the returning troops must have felt: "Two souls with but a single thought, two hearts that beat as one."

The back of this postcard reads, "Two soldiers at the Presidio, San Francisco, smitten by the bewitching damsel on the right agree to settle their differences by a boxing contest, while the object of their adoration stands intently watching her lovers."

An 1899 cartoon wash drawing shows a volunteer soldier just mustered out. The trooper cannot wait to strip off and throw away his uniform.

A memorial service in honor of the 1st Regiment of the California Volunteers, "the Fighting First," was held at the Alhambra Theatre for 85 soldiers from Northern California who died either from wounds or illness in the Philippines. Just 16 months earlier the 1st California had been encamped at Camp Merriam in the Presidio. Many veterans of this regiment are buried in the Presidio at the San Francisco National Cemetery.

The Presidio was overwhelmed by the large numbers of sick and wounded troops returning from Asia. It was necessary to build a larger more modern hospital. The first wing of the hospital was completed in June 1899, and in its first year the hospital cared for more than 5,000 patients. This view shows the new administration building.

Pres. William McKinley visited the Presidio on May 23, 1901, to officially dedicate the new hospital. The president was escorted by an honor guard of mounted San Francisco police officers as he addressed hundreds of soldiers and civilians. In this photograph by J.D. Givens, troops lift their hats to cheer the popular president. This was McKinley's last address to soldiers.

This view shows the completed U.S. General Hospital. Corridors and breezeways connected all the hospital buildings. The large house on the right was the hospital commander's quarters. In 1911, the hospital was named Letterman General Hospital in honor of Maj. Jonathan Letterman, an Army surgeon with the Army of the Potomac during the Civil War and later coroner for the city of San Francisco.

Less than five months later, he died from an assassin's bullet. (Courtesy of the National Park Service, Golden Gate Recreation Area, Park Archives and Records Center, Mary Conway Duryea Collection.)

COURT OF U. S. GENERAL HOSPITAL, PRESIDIO, SAN FRANCISCO

Lt. Col. George H. Torney became the hospital's commanding officer in 1904. He presided over the construction of an operating pavilion. The surgical ward and solarium were located in the center of a courtyard, shown in this J.D. Givens photograph. Lieutenant Colonel Torney stated that "the mission of the hospital was to develop a high standard of specialized professional services fitted to meet the demands of the Army."

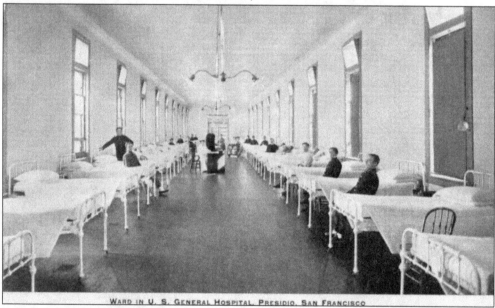

WARD IN U. S. GENERAL HOSPITAL, PRESIDIO, SAN FRANCISCO

Patients sit by their beds in this photograph of one of the wards. The ward was kept extremely clean and orderly in a military manner. Windows at frequent intervals provided natural light and window transoms allowed fresh air to circulate through the corridor. A nurse's station was located in the center of the ward. In 1905, the hospital staff consisted of 11 officers, over 150 Hospital Corps enlisted men, and 39 women in the Army Nurse Corps. This was the first Army general hospital to employ Army nurses, who were not considered commissioned officers.

Three

An Open Post

338 PRESIDIO AVE. ENTRANCE TO THE PRESIDIO. SAN FRANCISCO. CAL.

The Presidio Avenue Gate was the entrance to the post near Pacific Avenue. This 1908 postcard shows a pair of cannons with cannon balls atop brick gateposts. Two smaller upright cannons serve as barriers to protect the brick pillars. Beginning in the 1880s, this was the entrance favored by carriages and pedestrians en route to and from the city. Throughout most of its history, the Presidio of San Francisco was an "open post" where civilians could come and go onto the reservation without having to show passes or identification. The gates were usually left open and the guards would salute or acknowledge all visitors. Non-military visitors often drove or walked through the Presidio just to enjoy its park-like grounds. In April 1906, thousands of refugees from the earthquake and fire were given aid and assistance by the Army when they converged onto the post. During the Panama-Pacific International Exposition of 1915, one-third of the fair was located on the grounds of the Presidio, where thousands of visitors could partake in a wide variety of events and exhibits.

The brick guardhouse was erected at the north end of the main parade ground in 1899. Soldiers assigned to guard duty would gather at the guardhouse for inspection by the "officer of the day" before being posted and would remain there when not at their guard post. There were two large cages capable of holding 14 prisoners and six single cells were located in the basement. Serious offenders were sent to the military prison on Alcatraz. The guardhouse was later converted to a bank and a post office.

Board Walk, Presidio, San Francisco, Cal.

Presidio soldiers were assigned to roving patrols and guard posts. This solitary sentry patrols the elevated boardwalk on Lovers Lane.

Sgt. Edmond Jones, on the right, posts the guards at the Lombard Street Gate. While on guard duty these soldiers wore their dress blue uniforms and white gloves.

The Lombard Street Gate was the principal entrance to the Presidio by 1900. Visitors entering the post were met by guards, imposing sandstone pylons, an iron gate, and a sign warning that fast driving and racing were prohibited on the U.S. Military Reservation. Carved in the sandstone columns are the great seal of the United States and the War Department seal. Representations of "Liberty," "Victory" and the military symbols for the artillery, cavalry, engineer, and infantry branches are also prominently featured.

A 1901 postcard view shows a battery firing rifled 3.2-inch, breech-loading guns on the parade ground in front of the Montgomery Street barracks. In 1901, the Army reorganized the Corps of Artillery separating the coast artillery and the field artillery. In the next five years, the coast artillery at the Presidio increased from one company to ten, while the field artillery increased from one to three batteries.

Light Field Artillery drill at the Presidio is pictured on this 1901 postcard. The troops, ready for inspection, stand by their limbers and caissons, the two-wheeled horse-drawn vehicles and ammunition chests to which they would attach their guns for travel.

Cavalry troops practiced their horsemanship to develop their riding abilities. In one variation of "Roman riding," trios of horsemen stand and ride on the backs of four horses simultaneously. On this picture postcard mailed from the Presidio a soldier wrote, "This is what we call monkey drill. The cavalry does it all the time."

This photograph, titled "Final Round," shows a group of off-duty soldiers watching a pair of pugilists practice their fisticuffs. Each fighter has a second and the man behind the fighters in civilian clothes appears to be the referee. Boxing was a popular Army sport in the early days of the 20th century and different companies would compete with each other in the ring.

Pres. Theodore Roosevelt visited San Francisco on May 13, 1903. Much hoopla was made of the president's visit and city mayor Eugene Schmitz declared the day a holiday. Thousands of people turned out to greet the chief executive. For the first time in history, African-American troops were assigned as a presidential honor guard. Troops I and M of the segregated 9th Cavalry Regiment escorted the president from the Palace Hotel through the streets of San Francisco to the Presidio Golf Links. Maj. Joseph Garrard commanded the squadron. Capt. Charles Young, the only African-American West Point graduate on active duty,

commanded I Troop. The soldiers were resplendent, wearing their neat but simple blue uniforms with a pill box cap, white canvas leggings, and gloves. This photograph by J.D. Givens shows the president and his escorts arriving at the Presidio Golf Course. Six days later, Troops I and M, commanded by Captain Young, departed the Presidio on horseback for duty at Sequoia National Park. (Courtesy of the National Park Service, Golden Gate Recreation Area, Park Archives and Records Center.)

San Francisco was struck by an enormous earthquake on April 18, 1906. For three days fires raged through the most populated areas of the city. Over 200,000 refugees fled to safety in unburned areas, including the Presidio. The military, commanded by Brig. Gen. Frederick Funston, took a leading role in fighting the fires and providing assistance for those left homeless. The Army provided and pitched tents to protect thousands of refugees from the wet and cold weather. This postcard shows some of the thousands of refugees encamped in the Presidio.

Four permanent refugee camps were established within the grounds of the Presidio. These were part of the 30 official camps organized under the auspices of Maj. Gen. A.W. Greely. Camp One, located on the parade ground of the general hospital, is seen in this photograph by Arthur Clarence Pillsbury. The other Presidio sites were Camp Two, located in Tennessee Hollow,

This cartoon postcard by B.K. Leach shows refugees waiting patiently in the "Bread Line" for their rations. Food and other donations were sent from all over the country. The Army was responsible for the distribution of food to those in need. Distribution centers were established throughout the city.

Camp Three for Chinese at Fort Winfield Scott, and the Golf Links Camp Four. The camps were kept clean, neat, and sanitary. All of the camps provided common cooking and dining facilities, bathhouses, and toilets.

PRESIDIO CANTONMENT AND U. S. GENERAL HOSPITAL, SAN FRANCISCO

This is a view facing north of the Presidio in 1907. The U.S. General Hospital is located near the middle with the hospital parade ground, where refugee Camp One was located the year before, to the middle right. The barracks in the foreground are the East Cantonment, today's Ruger Street, Sherman Road, and Simonds Loop.

Officer's Quarters, Tennessee hollow. Presidio, Cal.

Officers' Quarters in Tennessee Hollow, in the Presidio's West Cantonment, were near the site of earthquake refugee Camp Two and the campsite of the Spanish-American War's Tennessee Volunteers.

42

ALAMEDA PRESIDIO CAL.

In 1860, the Army constructed the Alameda, a formal entry to the parade ground. By 1909, trees, plants, and cannon balls decorated the promenade in this view looking west toward the offices. On the right a soldier identified as "Jim" strolls with his three children. Children raised in military families are often known as "army brats," a nickname that is used even when the children are well behaved.

THE PARK AT THE PRESIDIO, S. F.

In this view looking east on the Alameda toward Presidio Boulevard, a sign reads "Visitors are prohibited from picking flowers or entering the yards of the officers' quarters."

The Army "Between Fires," a pun on a comic postcard, was mailed from the Presidio in 1909. During this era, the Army recruited only single men, and the low pay and lack of housing discouraged enlisted soldiers from marrying.

Senior noncommissioned officers did marry and were allowed to have their families live on post. Sergeants usually served several tours of duty before they could afford a wife, and families tended to be small. Soldiers often married the sisters and daughters of their fellow soldiers, ladies who were already familiar with Army life. Quarters were allotted based on rank and seniority, not on the size of the family. The 1909 brick two-bedroom duplexes on Riley Avenue were the first homes in the Presidio built especially for enlisted families.

The Presidio chapel was built in 1864 on the site of an earlier Spanish chapel. By 1910, the front of the church was covered with ivy. Both Roman Catholic and Protestant chaplains shared the building, conducting services, baptisms, weddings, and funerals. On Sundays there was a 9 a.m. Catholic mass, and at 11 a.m. the Protestant congregation would meet for worship. After 1909, Jewish soldiers were given permission to attend one of the synagogues in San Francisco. In 1931, the church became the Catholic Chapel of Our Lady and the new post chapel was used for Protestant services. The church was expanded and remodeled in 1952 and 1970.

The Ivy Covered Chapel, Presidio, San Francisco, California.

A Military Funeral
Presidio, S. F. Cal.

Deceased soldiers were entitled to a military funeral. An honor guard accompanied the flag-draped body as it was carried on a limber and caisson to the National Cemetery just west of the main post. In this picture, both officers and enlisted men pay their respects.

The Bachelor Officers' Quarters provided roomy accommodations for single officers or officers in transit. The imposing brick structure was built in 1903, replacing an earlier BOQ known as "the corral," which had burned in 1900. The wing on the left was named in honor of Maj. James Hardie and the right wing was named in honor of Capt. Erasmus D. Keyes. In later years, the building was named Pershing Hall in honor of General of the Armies, John J. Pershing.

Troops stand in formation for retreat in this 1911 photograph on the main parade ground. Buildings along the left side are the backs of the homes on officers' row, in the middle is the Bachelor Officers' Quarters, and to the right are the post chapel and two buildings used as

The Officers' Club was a dining facility and private club for officers. Also known as the Officers' Open Mess, the building was constructed on the remnants of the Spanish Comandancia adobe walls that were enclosed in lathe and plaster. Curb appeal was created by placing mortars and cannon in front of the building, as seen in this 1912 photograph by J.D. Givens.

either quarters or offices. The troops are facing west toward post headquarters and the flagpole, which is out of view.

Troops wait patiently as a field artillery battery forms for a road march. Horse-drawn 3.2-inch breech loading guns, limbers, and caissons carry the gun crews. Teams of six horses are needed to pull each fully loaded caisson and gun. More fortunate troops ride their own mounts. Wagons brought additional rations and supplies. Ambulance wagons, medical corpsmen, and a veterinarian also accompanied the battery to the field. The battery is formed up near the stables next to the guard house at the north end of the parade ground. (Courtesy of the National Park Service, Golden Gate National Recreation Area, Park Archives and Records Center, GOGA-1766.)

The field gun and crew are ready for action and watching for the enemy in this photograph of a sham battle. Sham or mock battles were opportunities for troops to train and practice their military skills in the field. The Army used sham battles to give demonstrations and provide entertainment at special events.

Cavalry troops on maneuvers set up this camp on the grounds of Fort Winfield Scott. On the right, troops slept in the smaller "pup" tents with their horses quartered nearby. A soldier writing home to Brooklyn, New York, in 1912 described the camp, "You can see the Signal Corps horses and camp on the extreme left. This is where we camped the three days we were out."

The cavalry barracks was constructed in 1912 and was the largest building on the post. It faced westward. Behind the barracks on the right is the post dispensary, the former Wright General Hospital. During World War I the building became the barracks and school for cooks and bakers. In 1919, the building was converted to offices and became the Ninth Corps headquarters. During World War II, it was the headquarters for the Western Defense Command and the Fourth Army.

Brig. Gen. John J. "Black Jack" Pershing, seen here with his staff in Texas, commanded the 8th Infantry Brigade at the Presidio for four months in 1914 before the brigade was ordered to the Texas-Mexico border. His family remained in their quarters at the Presidio (see page 18). On August 27, 1915, fire swept through the house, killing Frances Pershing and her three daughters, Helen, Ann, and Mary Margaret. Only six-year-old son Warren survived.

Street cars provided transportation to and from the Presidio and downtown San Francisco. The tracks crossed into the Presidio at Greenwich and Lyon Streets. In this 1915 photograph, the cars stopped at a terminal below the East (Presidio) Terrace officers' quarters. Car men and passengers could take advantage of the "Quick Lunch" stand. The "E" car line took over operation from the Presidio and Ferries Railroad in 1913 and lasted until June 1947. The "D" street car line operated from August 1914 to March 1950. The "G" Exposition line operated from February 1915 to September 1915.

In 1914, the Presidio prepared to take part in the Panama-Pacific International Exposition. Historian Frank Morton Todd described the plant nursery in the West Cantonment:

A permanent nursery was constructed in November of 1912 in Tennessee Hollow. The nursery consisted of six large green houses, with the necessary potting sheds, a heating plant and a large lath house. . . . thousands of small plants were removed, among them thousands of flats of acacia seedlings, eucalyptus seedlings, hydrangeas, marguerites, fuchsias, and other plants.

The world's fair was under construction in the Presidio in this 1914 view near Lyon Street. In the left foreground are the buildings of the Japan Pavilion. The tower of the Denmark Building is next, and then the Palace of Fine Arts colonnade and rotunda. The tower with the flag is the Cuba Pavilion. Other international pavilions of France, Denmark, Portugal, Italy, Bolivia, Sweden, Canada, and the Philippine Islands were along the Avenue of the Nations (Gorgas Avenue). Fourteen more nations were also represented by buildings in the Presidio. Before the

fair opened on February 20, 1915, it was decided that one of the requirements to be a guard was that the applicant must be a military veteran. On a postcard mailed to his brother from Angel Island, T.T. Busam, a soldier just completing his Philippine tour, wrote, "I may take a job as a guard at the fair if I pass the exam. Will get just a year's work out of it at 65 per and get to see the fair. Never did see a fair."

The Kansas State Building was one of the many state structures along the Avenue of the States and the Esplanade in an area that is now Mason Street at the east end of Crissy Field. Twenty-four states were represented by buildings in the Presidio. The model camp of the Marines, stockyards, dairy and poultry buildings, a wood-planked auto racetrack, and grandstands were located west of the state buildings.

A camp for military cadets participating at the fair was located on the north end of the parade ground near the cavalry barracks. In this postcard view mailed July 16, 1915, the Tower of Jewels can be seen on the far right. On the left past the barracks are the California State Building, the Oregon Building, and the tower of the Swedish Pavilion.

The Fort Point U.S. Life Saving Station was located at the far west end of the fair. The racetrack, which was also used as an athletic field, drill ground, polo field, and aviation field, can be seen in the foreground.

The Life Saving Service Station house was built in 1887, about 700 feet east of its present location. The house was an obstacle to the Exposition Company's plans to construct the planked racetrack. According to Frank Morton Todd, with everyone's agreement, the Exposition Company bore the cost of moving the station and building a new steel boat launch at the new site.

The Life Saving Station at Fort Point provided aids to navigation and rescue service along the coast and in San Francisco Bay. In 1914, the Life Saving Service merged with the Revenue Cutter Service and the new organization became what is now the United States Coast Guard. This unidentified Fort Point Station coast guardsmen wears his new uniform and in his right hand he holds a portable Kodak camera case. The photographs on the previous page and the next five pages came from his album.

A coast guardsman is seen here receiving instructions in semaphore signaling from a senior petty officer. He is giving the signal for "end of word." His postcard message was as follows: "The old man giving me hell; he says to tell you I am not in jail yet."

The U.S. Coast Guard Station Boat House, with a boat launch to the right, was built in 1914. The Coast Guard Station occupied 3.11 acres of land and 1.8 acres of tide and submerged land. In 1990, the Coast Guard moved to new facilities in Marin County and the station became part of the Golden Gate National Recreation Area.

The interior of the boat house was decorated with flags and banners. The 8,100-square-foot building was also used as a barracks and for offices. The main floor had six large rooms; six bedrooms and six latrines were located on the second floor.

The Coast Guard launches a boat using the marine railway from the boathouse in preparation for a lifeboat drill demonstration at the fair.

The six-man crew on the Fort Point Station launch rows into the exposition's Marina Harbor. The Navy cruiser in the background passes the breakwater and lighthouse. The 48 spotlights on the breakwater were manned by United States marines and provided the colored beams of light for the scintillator, which created the nightly light show at the fair. The lighthouse is still standing but is no longer functional.

In this William Foley photograph, the Coast Guard uses a yardarm to give a demonstration to a large crowd of spectators on the rescue and transfer of a passenger.

A lucky volunteer spectator is given a ride by the Coast Guard in one of the many special demonstration events at the exposition.

This photograph shows the racetrack's grandstands filled with thousands of spectators. The checkered flags remain from the International Grand Prix Automobile Race that took place a week after the opening of the exposition. The grandstands faced north with a view of the ships in San Francisco Bay.

The spectators came to see a spectacular event—a sham battle with the United States Navy. The "Blue Jackets" reenacted their successful landing and capture of Vera Cruz, Mexico, in April 1914. The Navy put on quite a show as they climbed over the barriers of the track, firing their rifles and dashing toward the enemy.

The "enemy" opposing the Navy assault was the Coast Artillery. Regiments of the 8th Infantry Brigade had left the Presidio with General Pershing for the Texas-Mexico border in pursuit of Pancho Villa, leaving only the Coast Artillery to maintain an Army presence. The caption on the postcard noted that the "Intrenched artillery, acting as infantry, routs army of invasion."

The Coast Artillery soldiers show off their infantry skills with a variety of positions in this demonstration of wall scaling.

VIATOR-ART-SMITH-IN-THE
MUZZLE-OF-A-12-INCH
MORTAR-FORT-WINFIELD
SCOTT-CALIF-4-22-
-15-P33-
©4-26-15-
C.G.COLLINS
-PHOTO-

Wearing his trademark cap worn backwards, daredevil aviator Art Smith visited the Coast
Artillery Corps at Fort Winfield Scott. In this gag picture by photographer C.G. Collins,
the diminutive Smith poses inside the muzzle of a 12-inch mortar. Smith was brought to the
Panama-Pacific International Exposition after native son Lincoln Beachey crashed his airplane
into San Francisco Bay. Always a crowd pleaser, the popular Smith, who was known as the
"Bird Boy," performed loops, death spirals, and often flew at night using phosphorous fireworks.
The exposition ended December 4, 1915. Smith continued to do stunt flying, performing the
following year in Japan. Even though he was a talented pilot, Smith was unable to enlist in the
Army Air Service during World War I. Some said it was because he failed to meet the height
requirement, others said it was because he had too many injuries from crashing his aircraft. The
Army did hire Smith to train new pilots for the war. Art Smith died in 1926 when he crashed
his U.S. Air Mail Service plane into a grove of trees.

Four

UNCLE SAM'S GUARDIANS
THE COAST ARTILLERY

One of Uncle Sam's Guardians, a 12-inch, breech-loading rifle pokes its muzzle over a concrete parapet towards the Pacific Ocean. The seacoast batteries at Fort Winfield Scott in the Presidio were part of a series of defensive coastal fortifications throughout the San Francisco Bay area. For nearly 60 years beginning in 1891, the 12-inch guns, 6-inch guns, and 12-inch mortars were the military guardians of the Golden Gate. The steel and reinforced concrete heavy-gun batteries were called "concrete soldiers," but as historian Brian B. Chin notes, "The army had prepared its artillery defenses at great effort without ever having the occasion to fire a shot against the enemy fleet that never showed up."

The Coast Artillery was established in 1901 when the U.S. Army decided to separate the light, mobile field artillery and heavy-fixed coast artillery into two separate branches. This illustrated postcard published in 1910 commemorates the new Coast Artillery Corps. An eagle perches above an American flag draped over the artillery symbol of crossed cannons on a scarlet oval; in the background is the breech side of a 12-inch gun; and in the foreground are a rifle, regimental crest, floating mine, and artillery shell. Two soldiers flank the symbols—the man on the left in denim-work fatigues and the soldier on the right in full dress blue uniform.

As the most permanent occupants of the Presidio, the Coast Artillery Corps troops were billeted in the large red brick Montgomery Street barracks. In this picture postcard, the 10th Company gathers on the barracks porch. The buildings were large enough for two companies to be comfortably housed with dining and kitchen facilities for both companies. In these barracks, the 10th Company is assigned the right wing and the 29th Company is assigned to the left wing.

The 10th Company of the Coast Artillery poses on the steps of their Montgomery Street barracks in this postcard dated 1909. The artillerymen are wearing uniforms with dark blue jackets with red chevrons, and light blue trousers with a red stripe on their pant legs. Their regulation hats are trimmed with a red band.

Carrying their rifles, field gear, and bedrolls, the Coast Artillery assembles on Montgomery Street in this postcard by photographer Charles Weidner titled, "Coast Artillery Going to Target Practice." Artillery soldiers were expected to know basic infantry skills and qualify as marksmen even though their primary duty was on the "big guns."

Coast Artillery troops stand at attention in this Charles Weidner photograph. A soldier's message reads, "This is the way we look when we go to camp or on the march." To perform their duty at the gun batteries, each day the artillery troops were required to march 1½ miles one way.

An unidentified coast artilleryman poses for his portrait on this patriotic postcard decorated with a drawing of cannon bursts, cannons, the American eagle, and the flag of the United States. Professional and amateur photographers often visited the Presidio to photograph the troops. The civilian post photographer James D. Givens provided photographic service to the Army for nearly 30 years. Some soldiers even had their own Kodak cameras, and many made "picture postcards." Photographic paper was available in the standard 5½ by 3½ postcard size and with preprinted postcard backs.

A comic postcard mailed from San Francisco in 1913, promoting C.H. Baker's shoe store, features the "Pelicannon, a strange bird you'll agree; In wartime he'd be useful in the coast artillery." It was a creation of popular *New York Herald* Upsie-Downs cartoonist Gustav Verbeck.

This 12-inch "disappearing rifle" is nearly hidden behind its concrete parapet. Dense plant growth in front of the battery provided necessary camouflage to completely hide the gun from view at sea. This allowed the gun crew to load and aim the weapon completely out of sight. The gun remained invisible until the artillerymen raised the muzzle over the crest of the parapet when ready to fire.

The gun crew prepares to load the 12-inch gun. The projectiles, weighing up to 1,000 pounds, required charges of over 200 pounds of powder. The guns required crews of up to 14 men, including at least 7 who just handled the ammunition. Shells and powder bags were maintained by the magazine crew in a protected bunker away from the gun. The powder was kept in long cylinder-like silk bags designed to fit into the gun's chamber. In this photograph, the breechblock is open as the crew prepares to load the powder bags.

Ready to fire, the 12-inch rifle makes its appearance above the parapet. The soldiers on the far right are working on the plotting board. The soldier on the far left is on the telephone with a direct line to the fire control station. Each Coastal Artillery gun had a gun director who was responsible for the maintenance and operation of the gun and for the safety and training of the gun's crew.

The 12-inch gun is seen here in action, actually showing the projectile as it propels out into the Pacific at an unseen target. This photograph by J.A. Wilson shows the awesome power of these weapons, which had a range up to 15 miles. Targets towed by civilian ships were used for practice fire. Actual firings were conducted only a few times a year. These were publicized in advance to warn fishermen to stay clear of the area. Nearby residents were advised to open their windows and doors since they could be damaged by the extraordinary concussions caused by the artillery fire.

"Do you recognize Captain Hilton?" was the message on this postcard. A captain commanded each battery. The executive officer, usually a first lieutenant, commanded the gun sections and for each gun in the battery there was a gun commander or gun director. The range officer was responsible for the plotting details and observers. The headquarters section of clerks, orderlies, cooks, and bugler was supervised by the first sergeant. A senior noncommissioned officer was in charge of the maintenance section.

Twelve Inch Guns In Action
Copyright J.A Wilson

The 12-inch rifle disappeared after firing when the resulting recoil knocked the gun back behind the parapet where the crew immediately began the process of swabbing out the gun and reloading. The gun crews worked rapidly. They were given training in maintenance and mechanics, and manual firing drills were constantly conducted to develop in the men an automatic sense of their specific duties in order to respond without verbal commands.

The gun crew and magazine crew at the end of the day pose on the disappearing carriage of their 12-inch rifle. The battery bugler, second from the left in the second row, is the only soldier wearing a campaign hat and not wearing denim fatigues. In order to serve in the Coast Artillery Corps these soldiers needed to be physically fit and meet all of the qualifications needed to enlist in the U.S. Army. For enlistment, the soldier was required to be at least 18 years old and not more than 35 years of age. He had to be at least 5 feet, 4 inches tall and weigh at least 115 pounds. The soldier could not be more then 6 feet, 6 inches tall or be considered obese. The applicant had to be certified to have good moral character by a reputable citizen. He had to pass a medical exam and have sufficient teeth to perform the function of mastication. All applicants were required to be able to read, write, and speak English, and to successfully pass the prescribed intelligence tests. Recruits had to be single, without dependents, and a citizen of the United States. In the segregated Army of this era, there were no African Americans in the Coast Artillery. Hispanics, Asians, and Native Americans could enlist in the Coast Artillery if they met all of the necessary qualifications. Once they enlisted, the soldiers were assigned to their regiment and then to their battery.

Soldiers used their fingers to plug their ears when firing 12-inch mortars at Battery McKinnon-Stotsenberg. Near the men are 12-inch shells on carriages ready for loading. The projectiles weighed 80 pounds, were breech-loaded, and came along with a silk bag of powder weighing up to 67 pounds. The minimum range of the mortars was 1 mile when fired at an angle of 70 degrees and a maximum range of nearly 7 miles when fired at a 45-degree angle.

Gunners in the pit at Battery McKinnon-Stotsenberg pose with their 12-inch mortars in the firing position. Four men imitate daredevil aviator Art Smith by climbing into the muzzle. A 12-man crew was needed to load and aim each mortar.

Fort Winfield Scott in the northwest corner of the Presidio was established as a separate Coast Artillery Corps fort on June 19, 1912, and was independent of the Presidio except for obtaining supplies. Coast Artillery troops formerly billeted on the main post relocated to their new barracks on the grounds of Fort Scott. In this photograph, a Coast Artillery regiment drills on their new parade ground in front of the band barracks.

The smallest but most ornate barracks was home to the Coast Artillery band. The Army traditionally provided band personnel with superior quarters befitting the talented musicians. The Fort Scott band barracks was built in 1912 in the Mission style with white walls and red roof tiles. Army prisoners could enjoy hearing band practice as the flat-roofed building to the left of the band barracks was the post stockade.

The Army hired civilian contractors to build barracks in the Mission style between 1908 and 1912. The barracks faced the horseshoe-shaped parade ground, located just a short walk from the Fort Scott batteries. This 1915 photograph by J.D. Givens shows the home of the 27th Company. The grounds have been landscaped, with lawn, flowers, and palm trees. The Army's favorite decorative element—artillery shells—grace the sidewalk in front of the building.

Men of the 27th Company of the Coast Artillery Corps are in their dress uniforms for this Christmas 1915 photograph. The three soldiers in the middle of the front row lead the company: the first sergeant, the captain commanding, and a lieutenant. In this era the Coast Artillery was the Army's most technological branch and officers were selected for their mathematical ability. Full strength for this company would be 105 men; the 27th Company was under strength with only 70 enlisted men.

Unlike the Mission-style enlisted barracks and offices of Fort Scott, the commanding officer's quarters on Upton Avenue was a red brick Georgian building with an expansive lawn and private walkway to the commander's nearby headquarters. The house is considered by many to be the most handsome home in the Presidio.

For recreation the officers and their families at Fort Scott could use the tennis courts and clubhouse seen in this J.D. Givens photograph. The windmill provided water for the clubhouse. The houses in the background are officers' duplexes on Kobbe Avenue.

Parades and inspections were a part of Fort Scott's daily routine. Wearing their winter overcoats, a Coast Artillery company stands at attention in front of their barracks.

The Coast Artillery open ranks for inspection in this 1917 photograph. With their new uniforms and field gear the troops prepare to go to war. During World War I, many coast artillerymen received abbreviated training before going overseas to serve in railway and field artillery batteries. The 67th Company from Fort Scott served with distinction in the Saint Mihiel and Meuse-Argonne offensives.

Five

WORLD WAR I

Liberty and the Flag
go well together in

SAN FRANCISCO, CAL.

Many of the regular Army troops vacated the Presidio in 1915 and 1916 to join "Black Jack" Pershing's punitive expedition into Mexico in pursuit of Pancho Villa. According to historian Erwin N. Thompson, the Presidio garrison shrunk to less than 100 men until Pres. Woodrow Wilson ordered 75,000 National Guard troops to join the regulars at the border and the Presidio became a receiving station for recruits enlisting in the service. The United States declared war on Germany on April 6, 1917, and officially became involved in the Great War to end all wars. With a great wave of patriotism sweeping the country, this postcard summed up the feelings of the people of San Francisco as the Presidio prepared for war.

In a postcard message sent in 1916, Miss Whitman wrote, "We visited this place the other day, this is where the soldiers stay, the different officers have their homes, the exposition was on part of these grounds, and it must have been beautiful when in its full swing, now it is over and they are wrecking the buildings fast." At the Presidio in 1917, on the former exposition grounds, the North Cantonment was quickly built to provide additional quarters and facilities for the incoming troops.

On the North Cantonment, single-story wooden barracks, mess halls, latrines, and storehouses were built for use by nearly 6,000 soldiers. The Sibley tents in the left foreground were provided for the overflow of troops. Buildings still remaining from the Panama-Pacific International Exposition, the Oregon Building, the Virginia Building, and the China Pavilion were put to use by the military. The Palace of Fine Arts was used as a warehouse and as a training facility.

78

The 12th Infantry left the Presidio with General Pershing in 1914 for the Mexican border. The regiment returned in May 1917 and was assigned to the North Cantonment. There the men of veteran 12th provided the cadre for two new regular Army regiments: the 62nd Infantry and the 63rd Infantry. This photograph shows the quarters of the 62nd Infantry. Regimental headquarters is the building on the far right with the staff car parked in front.

The Parthenon-like Oregon Building was located in the northeast corner of the North Cantonment and provided recreational facilities for the soldiers under the auspices of the Roman Catholic Knights of Columbus. On Saturday nights, dances were held beneath its lofty rafters. Many young women would come to dance with the men who were without friends in the city. The soldiers considered San Francisco to be the most hospitable city in the world.

63RD U. S. INFANTRY

PRESIDIO, SAN Fi

The 63rd U.S. Infantry Regiment was at full war strength when they stood on the main parade ground for this regimental portrait in August of 1917. The regiment commander, Colonel Richard C. Croxton, a West Point classmate of General Pershing, stands fourth from the left in the front row with his staff. The officers of the regiment stand in the front rows wearing their service hats; behind them stand the hatless enlisted troops. Combat ready for overseas duty with the 11th U.S. Division, these regulars never saw action in the Great War, remaining in reserve at the Presidio. As part of their training, the regiment heard a lecture by a female soldier,

R. C. CROXTON, COMMANDING

CALIFORNIA

COURSER, PHOTO, 2983 PINE STREET

Sgt. Ruth Farnam of the Serbian Army. She spoke about her experiences in the Balkan wars. Troops also attended rallies urging them to buy Liberty Bonds. During their stay at the Presidio, the soldiers were permitted to go off post after duty hours. No passes were ever required, there was no bed check at taps, and the only requirement was that the soldiers be well behaved and return by reveille the next morning. Cases of AWOL were rare. Only in September–October of 1918 were the troops restricted to the Presidio when the worldwide influenza epidemic struck San Francisco.

The Presidio Motor Transportation Corps, the 6th Motor Command, stand by their trucks in the North Cantonment ready to be inspected. Note the two-man motorcycle with a sidecar on the left. The 6th Motor Command was made up of the 406th and 411th Motor Supply Trains. Prior to World War I, transportation operations were the responsibility of the quartermaster corps.

During the Mexican Punitive Expedition there was a limited use of motor cars. World War I would be the first conflict to make full use of motorized vehicles. Following the war, the Motor Transportation Corps tested their capabilities in November 1919 with a successful transcontinental road march on the Lincoln Highway from Washington, D.C., to the Presidio of San Francisco.

In this April 27, 1918 photograph taken in the North Cantonment, members of the supply company of the 62nd U.S. Regiment wash their tin cups and plates after eating. In a few months the troops assigned to the 8th U.S. Division would be on their way to Russia with the American Expeditionary Force Siberia.

Fort Winfield Scott had an active training program during the war. The parade ground provided a campsite for these troops being inspected while they stand at attention in front of their "dog" or "pup" tents—two shelter halves buttoned together and shared by two men.

The first officer training school opened at the Presidio in May 15, 1917, for 2,500 enlisted men and civilian recruits. At the end of the three-month course, over 1,000 officer candidates were recommended for reserve commissions. A second officer training camp began immediately in August 1917. Companies of officer candidates are lined up for the evening parade in this July 1917 photograph.

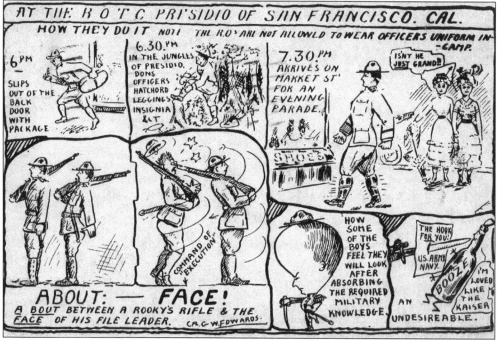

"About: Face! How They Do It," explains the experience of being a reserve officer candidate at the Presidio, as documented in a series of cartoons by Corp. G.W. Edwards.

"Storming 'Commission' Ridge" for the candidates involved three months of intense physical and classroom training, as seen in this Corporal Edwards illustration. In the first class, only 40 percent of the candidates received commissions.

This Corp. G.W. Edwards comic postcard puns "close order, extended order," while the cadet sleeps in on a Sunday morning and dreams of charging an imaginary enemy who looks just like the Kaiser. But for the officer candidates, the training was serious. Candidate Clayton Taylor of ROTC Company "8" wrote his brother, "The San Francisco spirit is great. My only aim now—is to make good and do my bit."

Officer candidates march on Lincoln Boulevard past Letterman General Hospital. Designated the ROTC Provisional Regiment, only 250 college students participated in the third three-month reserve officer training class, which began operations in July of 1918. The students represented colleges and universities from all of the Western states.

The San Francisco municipal railway ran two streetcar lines to the Presidio from the Ferry Building loop: the "D" Geary–Van Ness and the "E" Union Street. A new covered shelter was provided for waiting passengers at the street car terminus near Letterman Drive and Presidio Boulevard.

Interested in the welfare of service men, the YMCA purchased the Enlisted Men's Club at the conclusion of the Panama-Pacific International Exposition. Originally located near the Canada Pavilion, the building was moved to a new site on Lincoln Boulevard opposite Letterman Hospital. A reading room, pool tables, piano, phonograph, grill, and canteen were located on the main floor. A gymnasium was on the lower floor. A 900-seat auditorium on the third floor provided movies and live entertainment including a production of Gilbert and Sullivan's *Mikado*.

The National Defenders Club was located just outside the Lombard Street Gate. A canteen, library, billiards, and dances every night except Sunday were the club's attractions. Essentially a soldier's club, sailors were welcome, as seen in this photograph. A favorite activity was when the Navy challenged the Army to a singing match.

Letterman General Hospital and Y.M.C.A. at The Presidio. of S.F. 1918.

Despite the war in Europe, the grounds between the YMCA and Letterman General Hospital in 1918 appear quiet and idyllic. In the two years of American involvement, the hospital admitted 18,700 patients. The war ended November 11, 1918. The following year, 12,400 sick and wounded soldiers arrived from Europe. The hospital had an orthopedic center for amputation cases, neurology and psychiatric clinics, and a venereal disease clinic.

AEROPLANE VIEW OF LETTERMAN GENERAL HOSPITAL
PRESIDIO, SHOWING PALACE OF FINE ARTS
SAN FRANCISCO, CALIF.

Letterman General Hospital is viewed from the air in this 1919 postcard view. The North Cantonment is in the upper left and the Palace of Fine Arts and lagoon are in the upper right. The lagoon was located just outside the boundaries of the Presidio. On the right are barracks and wards built for use by the hospital. In 1927, after considerable negotiations, the U.S. Army transferred the Palace of Fine Arts to the City of San Francisco for the right to run the Belt Line Railway from Fort Mason to the Presidio via Marina Boulevard.

Six

CRISSY FIELD

Crissy Field opened July 1, 1921, as part of Northern California's coastal defenses. The Air Coast Defense Station was named by Maj. Henry H. "Hap" Arnold in honor of his friend and colleague Maj. Dana H. Crissy, who died when his plane crashed on October 8, 1919, while participating in the U.S. Army's Transcontinental Reliability and Endurance Test. The 91st Observation Squadron's Maj. George H. Brett was the field's first commander. Crissy Field can be seen in this c. 1930 aerial view of the northwestern corner of the Presidio.

This is Crissy Field as it appeared in the era of its first two commanders Maj. George H. Brett and Maj. Delos C. Emmons. Both men would be general officers in World War II. Brett would be U.S. air commander in the southwest Pacific until July 1942 and Emmons would command the Hawaiian Department after Pearl Harbor and later the Western Defense Command at the Presidio. This photograph was taken about 1925 from Crissy Field Avenue on the bluffs at the west end of the field. The building in the center foreground is the garage. The flagpole stands between the garage and the administration building, headquarters for Crissy Field. The large Mission-style building is the enlisted men's barracks. Facing the landing field are the two reserve hangars. The word "Crissy" is painted on the roof of one and the word "Field" is painted on the roof of the other. Next is the U.S. Air Mail hangar and in the distance lies the motor transport buildings. The North Cantonment had been cleared of buildings. The dome of the Palace of Fine Arts and the city of San Francisco are visible in the background.

A pilot stands here with his Loening OA-1A, an amphibious observation airplane powered by an inverted Liberty engine. According to National Park Service historian Stephen A. Haller, the Army reported a pair of Loening amphibians at the field in 1925. In cooperation with the Navy for a non-stop flight to Hawaii, the Army cleared a previously unused seaplane ramp. The Army pilots chose to land and take off their OA-1A's from the airfield, not the seaplane ramp.

Army pilot Sgt. D.A. Templeton, wearing his regulation coveralls, stands by his airplane, a DeHavilland-4 (DH-4). The flying sergeants were the unheralded pioneers of aviation. Enlisted men were allowed to earn their pilot's wings as a result of shortages of qualified officer pilots. Enlisted pilot training was informal, with the sergeants usually learning from an experienced pilot in their squadron. The Congressional Air Corps Act of 1926 made their training official. The opportunity for enlisted men to serve as pilots ended in 1942 with the Flight Officer Act.

Along a dirt road (Mason Street), two of the 91st Observation Squadron's enlisted ground crew sit in their gasoline fuel tank truck waiting to refuel arriving aircraft. The vehicle shows the U.S. Army Air Corps insignia—a white star with a red disk within a blue disk.

A U.S. Army C-2 dirigible visited Crissy Field on September 27, 1922, as the final stop of a transcontinental flight from Virginia. Built by the Goodyear Rubber Company, the airship required a hangar measuring 1,000 feet in length with a 125-foot clear span. Thousands of visitors came out to view the silent behemoth. The C-2, used primarily for observation, took a group of officers from Ninth Corps Area Headquarters on an airborne tour of San Francisco and the Bay Area.

"THE HOP-OFF"
...1 the world Flyers Leaving Crissey Field, San Francisco, Sept. 27, 192
R.W.D.

On September 25, 1924, the "Round the World" International Competition came to the Presidio where three U.S. Army Douglas World Cruiser aircraft "hopped off" at Crissy Field. Four planes named for the cities of Chicago, New Orleans, Seattle, and Boston departed Seattle for a round-the-world flight. The *Boston* and the *Seattle* crashed, but the remaining two aircraft successfully circled the globe. The *Chicago* and *New Orleans* accompanied by a replacement plane, the *Boston II*, received a rousing welcome before proceeding to Seattle, Washington, to complete the 26,345-mile journey.

An imposing barracks for the enlisted men assigned to Crissy Field was completed in 1921. Constructed in the Mission style with red roof tile and white painted concrete, the three-story building with open verandas and its own post exchange was home to most of the 198 enlisted men assigned to the 91st Observation Squadron and 15th Photo Section. The building was later converted to offices and in 1946 was named Stilwell Hall in honor of Gen. Joseph W. Stilwell.

The spacious barracks included a pleasant dayroom where off-duty enlisted men could relax. The dayroom included a library, piano, tables for writing, and comfy chairs. Furniture was provided by the post quartermaster, and books and magazines were paid for by the unit fund.

The barracks verandas were located on two stories on the north, east, and west ends of the building. Originally open, the porch was enclosed and furnished with deck chairs and a swing. The men could use the veranda as a sun parlor, depending on the time of day, and lamps were provided for evening use.

Fine dining was available in the squadron mess hall. Troops served themselves and sat on functional, but not very comfortable, stools. Instead of duty as kitchen police (KPs), at least two lucky troops would serve each day as dining room orderlies responsible for keeping the area clean.

The kitchen was kept spotless and KPs were expected to scrub and polish every pot, pan, boiler, and coffee urn after each meal. Garbage cans were lined with newspapers to keep the containers from becoming greasy. "Good cooks are difficult to find outside of the army," was a favorite saying. The 1926 Christmas menu for the 91st Observation Squadron included roast turkey, roast leg of lamb, shrimp salad, mashed potatoes, baked potatoes, fig bars, lady fingers, and strawberry Jell-O with whipped cream.

In 1925, the Army began replacing the older, worn-out DeHavilland-4s with newer aircraft, the Douglas O-2H, Douglas O-25C, and later the Boeing P-12. The 91st Observation Squadron insignia can be seen on the fuselage of this Douglas O-25C as it prepares to take off from Crissy Field.

An Army airplane goes down in San Francisco Bay near Crissy Field in this photograph by the author's father, Robert Bowen Sr., taken about 1930. In its first ten years, Crissy Field averaged eight crashes per year. The first fatality, researched by historian Stephen A. Haller, was "a reserve officer on one year's active duty status . . . crashed in the Bay . . . while attempting a take-off . . . in a heavy wind the ship suddenly went into a spin from an altitude of about 150 feet."

The Douglas BT-28 seen at Crissy Field in this December 21, 1930, photograph was used primarily as an Army Air Corps Basic Trainer for new pilots. Squadrons of the Organized Reserve Corps were based at Crissy Field. Except while on active status, reserve officers were not paid and flying was the only incentive for membership in the Air Reserve.

During the 1920s and 1930s, non-military aircraft frequently landed at Crissy Field. This airplane was decorated with stars and stripes. According to historian Stephen A. Haller, Crissy Field played host to 244 commercial aircraft in 1931. That same year, there was a record of no less than 812 visiting military airplanes. None of these visits was in connection with the Army's annual maneuvers.

A squadron of six bombers flies over Crissy Field on the afternoon of February 9, 1933. Despite budget constraints and a conservative Army leadership slow to accept new concepts in air power and aircraft design, the airplanes landing and taking off from Crissy Field were getting larger and faster. The Keystone bombers seen in this photograph were among the last of the biplane bombers used by the Air Corps. The appearance and performance of all metal monoplane bombers in 1934 would make biplane bombers obsolete. Crissy Field, which once had been known as "the last word in airfields," was becoming obsolete, unable to expand its runway for larger and faster aircraft. According to Erwin N. Thompson, in 1933 there were 23 aircraft permanently based at Crissy Field: observation planes, a transport plane, a photographic plane, and a trainer. Two years later, when the War Department began the process of closing the airfield, only five airplanes were based there. In 1935, the U.S. Army Air Corps opened Hamilton Field, a larger, more expansive air base north of San Francisco in Marin County. (Courtesy of the San Francisco History Center, San Francisco Public Library.)

Seven

PEACETIME ARMY
1920s–1930s

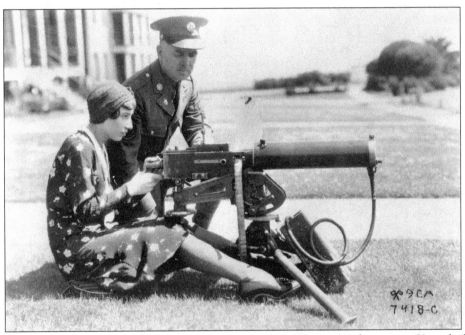

"Take your daughter to work day?" might be the perfect caption for this image. Years before women had an opportunity to learn basic combat skills in the U.S. Army, First Sergeant Shramar of the 30th Infantry Regiment taught his daughter how to use a Browning M1917A1 .30-caliber, water-cooled machine gun. This family portrait was taken in front of the Montgomery Street barracks. Following World War I, Congress authorized the Army to be reduced in size and strength to less than 119,000 men. Military historian Russell Weigley considered the peacetime Army of the Roaring Twenties and Depression Thirties less ready to function as a fighting force than at any time in its history. For the small cadre of Army "regulars" and their families, being stationed at the Presidio of San Francisco was considered to be a great duty station. (Courtesy of the National Park Service, Golden Gate Recreation Area, Park Archives and Records Center, GOGA-1766.)

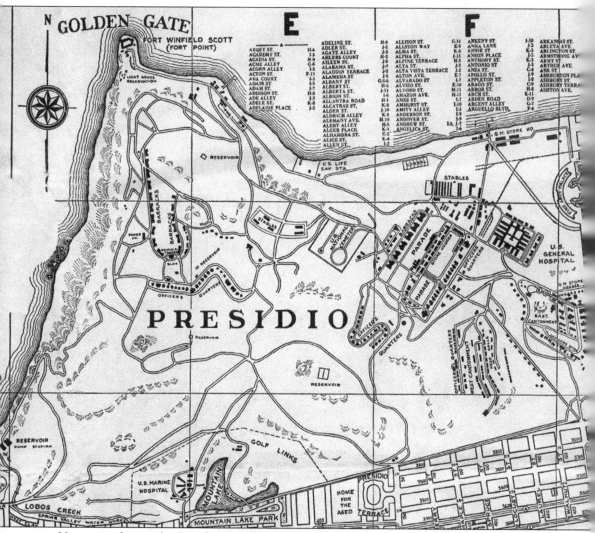

Visitors exploring the Presidio in 1926, whether walking or driving, could have easily found their way anywhere on the reservation using this detailed Thomas Brothers map. Paved roads, reservoirs, numerous buildings, including two hospital complexes, and the golf links show the improvements the Army made in the Presidio over 75 years of occupation. Although their locations were well known and many were disarmed, Coast Artillery gun emplacements were considered classified information and are not shown on the map. Crissy Field was not shown, although its existence was certainly no secret. During the isolationist 1920s and 1930s, the Presidio was often threatened by proposals to change its military use or close it down. One proposal was to convert the entire post into a giant military hospital. In 1927, local real estate interests launched a campaign to have the Army sell the reservation so the land could be subdivided. The Army chose to continue to keep its military presence in San Francisco and not close the Presidio.

The "Garden Spot of the Ninth Corps" can be seen in this 15th Photo Section aerial view of the Presidio's main post in 1925. The 30th Infantry Regiment, "San Francisco's Own," prepares to march on the old main parade ground. The 30th Infantry was still using mule- or horse-drawn wagons for the march. Directly below the parade ground are a tennis court and baseball diamond. To the left, divided by the Alameda, are the houses on officers' row, now named Funston Avenue in honor of the 1906 hero. At the top of the parade ground on Moraga Street are the Pershing Hall BOQ, the chapel, quarters, and officers' club. To the right of the parade ground is the bandstand where the old commander's quarters, site of the Pershing family tragedy, was located. The square would be named Pershing Square in honor of the general and his family. At the lower right are the Civil War–era barracks converted to headquarters offices. Near the top center of the photograph, along the curving ridge, are the duplex homes on Infantry Terrace, which were quarters for the officers of the 30th Infantry Regiment between 1922 and 1941. (Courtesy of the National Park Service, Golden Gate Recreation Area, Park Archives and Records Center.)

CAR STATION
PRESIDIO of SAN FRANCISCO, CALIF.

A military policeman was responsible for traffic control at Letterman Drive and Presidio Boulevard. A new streetcar station was constructed in 1920 to replace the open shelter and lunch stand of World War I. The building included a waiting area, a post exchange, and a grill. The fare for the 30-minute trip to downtown San Francisco was 10¢. Taxis at a nearby taxi stand charged about $1.25 for the trip downtown.

30TH U.S. INFANTRY BAND - PRESIDIO
SAN FRANCISCO, CALIF.

A tall drum major with a white bearskin hat leads "San Francisco's Own" 30th U.S. Infantry Regimental Band on parade down Lincoln Boulevard. The standards for regimental bandsmen were high. In order to enlist in the band, talented musicians had to be able to read music as well as perform their basic soldierly skills. Bandsmen also provided music for concerts, recitals, military balls, and dances.

Polo was the favorite sport of many Army officers. In the 1920s and 1930s, there was a Presidio polo team and mounts were always available. A 900-foot-long polo field was set up at the east end of Crissy Field. This 1930 Presidio Polo Club program lists the season's military and civilian opponents: the 11th Cavalry, Presidio of Monterey, Burlingame–San Mateo, Oakland, Napa, and Menlo Park Polo Clubs. Polo-enthusiastic commanders would sometimes gather crack polo players for their staff. One critic of polo, Maj. Gen. Johnson Hagood, felt there was no military value in the game and that polo, including grooms, stables, and remount stations, cost the taxpayers too much at $100,000.

Presidio Polo Club

Summer Season 1930

All Games Broadcast Over KJBS by Flynn and Collins, Ford Dealers
Van Ness Ave. at Sacramento St.

Fishing was a popular pastime for many soldiers, who took advantage of the piers and beaches of the Presidio to cast their lines. Warrant Officer Frank L. Bowen stands beside the catch of the day behind his West Cantonment quarters. In 1932, the Army demolished a group of 19th-century wood-frame barracks in the West Cantonment (Portola Street) and replaced them with new Georgian Revival brick duplexes for married warrant officers and senior noncommissioned officers.

The Chapel — Presidio of San Francisco, Calif. — Ribl [?]

A new post chapel for Protestant services was dedicated in 1931. The Spanish Colonial–revival building, with a bell tower, is located just east of the national cemetery on a bluff overlooking the Presidio and San Francisco Bay. The side entry stairs in this view of the chapel lead directly to Victor Mikail Arnautoff's fresco mural, *The Peacetime Activities of the Army*. Artist Willemia Muller Ogterop was commissioned to design the stained-glass windows to enhance the beauty of the chapel.

One of the scenes in Victor Arnautoff's mural depicts the Army's involvement in the planning of the Golden Gate Bridge. A grimacing officer listens to a civilian engineer explain the bridge's structure. The War Department had to give permission to the Bridging the Golden Gate Association before they could proceed with the project. Military men worried that enemy bombing could easily destroy the bridge and bottle up San Francisco Bay.

104

Ninth Corps' Commanding Gen. Malin Craig opened Crissy Field to the public for a spectacular Army Day celebration. The event featured a parade by the Provisional Brigade, a formal guard mount, tactical exercises, and the ever popular sham battle. The Junior Chamber of Commerce sponsored Army Day, and San Francisco supervisor Jesse C. Colman chaired the Citizens' Committee.

Photographed from Battery Sherwood, the west end of Crissy Field can been seen in this 1935 photograph of the Golden Gate Bridge during construction. Most of the aircraft had been moved to Hamilton Field. Despite their initial objections, the military allowed Presidio access. Chief Engineer Joseph B. Strauss established a field office at old Fort Point for his resident engineer's headquarters and construction began. Two elevated highways (Doyle Drive and Park Presidio State Highway One) crossed through the Presidio to provide approaches to the new bridge.

Presidio Officers Club.
"Oldest building in San Francisco."

The Presidio Officers' Club was known as the oldest building in San Francisco, even though little of the original adobe was visible and nothing remained of the original roof. A Federal and Local Works Administration project in 1934 "restored" the building. The restoration was actually a remodel that replaced the existing wood walls and shingled roof with red tile and white stucco to make it look more Spanish.

MAC ARTHUR AVENUE
LOVERS LANE
WEST CANTONMENT

ALL CANTONMENT QUARTERS.
EXCEPT NOS. 790 to 800 Inclusive
TO WHICH MAY BE HAD VIA PRESIDIO
UP HILL FROM CAR STATION

In this 1938 photo, the author's mother, Lucy Bowen, newly arrived from Leavenworth, Kansas, stands by the directional sign located near Presidio Boulevard and MacArthur Avenue. West Cantonment would later be renamed Portola Avenue. Nearby Lovers Lane, in the Tennessee Hollow with a path crossing a stone footbridge over the Central Creek tributary, was considered by many to be the most romantic spot to stroll.

From May 27 to June 2, 1937, the Golden Gate Bridge
Fiesta celebrated the opening of the Golden Gate Bridge. It
began with the "Opening Day Pedestrian Bridge Walk" and
a "Spectacular Day Parade" to Crissy Field, where the crack
11th Cavalry from the Presidio of Monterey performed. San
Franciscans partied for a week.

The presidential limousine of Franklin Delano Roosevelt cruises along the east end of Doyle
Drive after the President's visit to the Golden Gate Bridge in July 1938. The 30th Infantry
Regimental Band plays and the regiment stands at attention near the colonnade of the Palace
of Fine Arts in this photograph by the author's father, Robert W. Bowen Sr.

Sgt. David Gillimer gives three students from San Francisco's Argonne School Junior Traffic Patrol a demonstration on an anti-aircraft gun. The city's student traffic patrol was rewarded with a special day at the Presidio each year. The young men (there were no girls in the traffic patrol) were treated as honored guests by the Army. The boys in this October 28, 1938, *San Francisco News-Call Bulletin* photograph would be old enough to serve in the Army in the later days of World War II. (Courtesy of the San Francisco History Center, San Francisco Public Library.)

Eight

WORLD WAR II

Amassed and ready to defend the Bay Area within hours of Japan's December 7, 1941 attack on Pearl Harbor, thousands of soldiers assemble at Crissy Field. Coastal Harbor defenses were already on full alert. For San Franciscans, the Presidio would be the first line of defense. Lt. Gen. John L. DeWitt, commander of the Fourth U.S. Army and the Western Defense Command, headquartered at the Presidio, declared the West Coast a wartime "Theatre of Operations" and considered it a combat zone. In the next four years, the Presidio would be the major hub of wartime operations in the western United States. (Courtesy of the National Park Service, Golden Gate Recreation Area, Park Archives and Records Center.)

On November 1, 1941, in anticipation of war with Japan, the U.S. Army established the Military Intelligence School to study the Japanese language. Attending the school were 60 students, including 58 Americans of Japanese ancestry. School commander Lt. Col. John Weckerling, assisted by Capt. Kai Rasmussen, selected a talented faculty of nine expert Japanese linguists. The school and student billet was located in the old U.S. Air Mail hangar. Building 640 is seen in this photograph as it appears today.

The course of study for the Intelligence School attendees was demanding. Students seen in this photograph were trained in social and cultural traditions, translation, and battlefield interrogation techniques. Forty-four students successfully completed the course and graduated in April 1942. The following month the school was moved to Camp Savage, Minnesota. Instructor Shigeya Kihara described the experience in 1941: "Heretofore, Japanese Americans were considered second-class citizens, linked to Japan and not to be trusted. Here they were asked to do something of vital service to the United States, very critical not only to the U.S. Army but for Japanese Americans." (Courtesy of the National Park Service, Golden Gate Recreation Area, Park Archives and Records Administration.)

Presidio of San Francisco, California
May 3, 1942

INSTRUCTIONS
TO ALL PERSONS OF
JAPANESE
ANCESTRY
Living in the Following Area:

All of that portion of the City of Los Angeles, State of California, within that boundary beginning at the point at which North Figueroa Street meets a line following the middle of the Los Angeles River; thence southerly and following the said line to East First Street; thence westerly on East First Street to Alameda Street; thence southerly on Alameda Street to East Third Street; thence northwesterly on East Third Street to Main Street; thence northerly on Main Street to First Street; thence northwesterly on First Street to Figueroa Street; thence northeasterly on Figueroa Street to the point of beginning.

Pursuant to the provisions of Civilian Exclusion Order No. 33, this Headquarters, dated May 3, 1942, all persons of Japanese ancestry, both alien and non-alien, will be evacuated from the above area by 12 o'clock noon, P. W. T., Saturday, May 9, 1942.

No Japanese person living in the above area will be permitted to change residence after 12 o'clock noon, P. W. T., Sunday, May 3, 1942, without obtaining special permission from the representative of the Commanding General, Southern California Sector, at the Civil Control Station located at:

Japanese Union Church,
120 North San Pedro Street,
Los Angeles, California.

Such permits will only be granted for the purpose of uniting members of a family, or in cases of grave emergency.

The Civil Control Station is equipped to assist the Japanese population affected by this evacuation in the following ways:

1. Give advice and instructions on the evacuation.
2. Provide services with respect to the management, leasing, sale, storage or other disposition of most kinds of property, such as real estate, business and professional equipment, household goods, boats, automobiles and livestock.
3. Provide temporary residence elsewhere for all Japanese in family groups.
4. Transport persons and a limited amount of clothing and equipment to their new residence.

The Following Instructions Must Be Observed:

1. A responsible member of each family, preferably the head of the family, or the person in whose name most of the property is held, and each individual living alone, will report to the Civil Control Station to receive further instructions. This must be done between 8:00 A. M. and 5:00 P. M. on Monday, May 4, 1942, or between 8:00 A. M. and 5:00 P. M. on Tuesday, May 5, 1942.
2. Evacuees must carry with them on departure for the Assembly Center, the following property:
   (a) Bedding and linens (no mattress) for each member of the family;
   (b) Toilet articles for each member of the family;
   (c) Extra clothing for each member of the family;
   (d) Sufficient knives, forks, spoons, plates, bowls and cups for each member of the family;
   (e) Essential personal effects for each member of the family.

All items carried will be securely packaged, tied and plainly marked with the name of the owner and numbered in accordance with instructions obtained at the Civil Control Station. The size and number of packages is limited to that which can be carried by the individual or family group.

3. No pets of any kind will be permitted.
4. No personal items and no household goods will be shipped to the Assembly Center.
5. The United States Government through its agencies will provide for the storage, at the sole risk of the owner, of the more substantial household items, such as iceboxes, washing machines, pianos and other heavy furniture. Cooking utensils and other small items will be accepted for storage if crated, packed and plainly marked with the name and address of the owner. Only one name and address will be used by a given family.
6. Each family, and individual living alone, will be furnished transportation to the Assembly Center or will be authorized to travel by private automobile in a supervised group. All instructions pertaining to the movement will be obtained at the Civil Control Station.

Go to the Civil Control Station between the hours of 8:00 A.M. and 5:00 P.M., Monday, May 4, 1942, or between the hours of 8:00 A.M. and 5:00 P.M., Tuesday, May 5, 1942, to receive further instructions.

J. L. DeWITT
Lieutenant General, U. S. Army
Commanding

In the spring of 1942, Lieutenant General DeWitt and the West Coast Defense Command, headquartered at the Presidio, enforced Pres. Franklin Roosevelt's Executive Order 9066. This resulted in removing from the West Coast and the incarceration in camps of over 110,000 people of Japanese ancestry—more than two-thirds of those born in the United States. These citizens were not individually charged, but were collectively ordered to report for internment, as seen in this poster used in Los Angeles, California. Instructions authorized by General DeWitt were also posted in San Francisco, Oakland, Sacramento, and communities with large populations of Japanese Americans and resident aliens. The Army interned some German and Italian resident aliens of questionable loyalty, but no other United States citizens were targeted for persecution and deprived of their rights of due process because of their national origins. When reporting, families could bring only what they could carry. They were not even allowed to bring their pets. During the war, the West Coast Defense Command Headquarters was located in Building 35 (see page 50), the old cavalry barracks and Cooks and Bakers School.

111

The Women's Army Auxiliary Corps (WAAC), commemorated on this postcard, was established by an Act of Congress in 1941. The WAAC and the Women's Army Corps (WAC), as they were known after 1943, provided non-combat support in offices, labs, and motor pools. WAC companies were stationed at the Presidio in headquarters and Letterman Hospital. To provide housing for the women, temporary wood barracks were built near Crissy Field and Funston Field (Moscone Playground), a city park leased to the Army. The Greek goddess Athena was used as the symbolic insignia of the WAC.

Battle casualties and tropical disease cases from the Pacific arrived on hospital ships at Fort Mason. In this 1942 photograph, a convoy of Army ambulances from Fort Mason en route to Letterman General Hospital enters the Presidio through the Lombard Gate. During the war the Presidio was no longer open to the general public, as the sign in front of the new sentry box advises visitors to "enter on business only." (Courtesy of the National Park Service, Golden Gate Recreation Area, Park Archives and Records Center, GOGA-2991.)

Letterman General Hospital served triple functions as a general hospital, embarkation hospital, and evacuation hospital. In 1944, with increased war activity in the Pacific, Letterman's primary purpose was receiving overseas patients and promptly evacuating them to other hospitals near their homes. Hospital trains using the Belt Line track at Crissy Field were the primary means of transporting patients. (Courtesy of the National Park Service, Golden Gate Recreation Area, Park Archives and Records Center, GOGA-2266.)

A patient was loaded onto a hospital ward car and accompanied by an enlisted escort. Escorts were permanently assigned to the Hospital Train Unit (Service Command Unit 1960) and would return to Letterman to escort another patient; as many as four full trains operated out of the Crissy Field spur every day. Each train had 10 to 12 cars. A ward car had eight two-tier bunks. There were also cars for orderlies and a combination kitchen-dining-pharmacy car. An officer car had facilities for four medical officers and six nurses. (Courtesy of the National Park Service, Golden Gate Recreation Area, Park Archives and Records Center, GOGA-1766.)

Wearing their "Letterman egs," patients pose on the hospital's courtyard lawn. Letterman was one of the nation's foremost hospitals in prosthetics and rehabilitation of amputees. Major divisions in the hospital included surgery, cardiology, neurology, radiology, urology, dental, and outpatient care. In 1940, there were 9,064 admissions; from 1941 to 1946 the hospital admitted a total of 193,429 patients, reaching a high of 73,452 in 1945. In 1945, hundreds of people lined Lyon Street to greet the "Angels of Bataan" on their way to the hospital. The "Angels" were Army nurses who had been held as prisoners-of-war in the Philippines for nearly four years. (Courtesy of the National Park Service, Golden Gate Recreation Area, Park Archives and Records Center, GOGA-35288.)

The West Coast Memorial to the Missing of World War II was dedicated on November 29, 1960. The monument on the bluffs of Fort Winfield Scott overlooking the Pacific Ocean list the names of 413 members of the armed forces who were lost or buried at sea during the war. The memorial was designed by the San Francisco architectural firm of Clark and Beuttler. Lawrence Halprin designed the landscape; Jean deMarco sculpted the statue of Columbia.

Nine

SIXTH U.S. ARMY
1946–1994

In a major reorganization of the Army in 1946, the Sixth U.S. Army, of World War II fame in the Pacific Theater, was reactivated as one of six U.S. Continental Armies. The Sixth Army, commanded by Gen. Joseph W. "Vinegar Joe" Stilwell, replaced the Fourth Army. A pair of large buildings originally built as barracks in 1940 served as the headquarters in the Presidio. The Sixth U.S. Army was now responsible for the ground defense of the western third of the continental United States in an area that extended from the Pacific Coast to the Rocky Mountains.

CUSTODIAL SERVICE BRANCH
LETTERMAN GENERAL HOSPITAL, PRESIDIO OF S

FRANCISCO, CALIF

POST PHOTOGRAPHER
PRES OF S.F.
SEPT 1949
11-A

Declaring that "there shall be equality of treatment and opportunity for all persons in the armed services without regard to race, color, religion or national origin," Pres. Harry S. Truman issued Executive Order 9981 on July 26, 1948. This began the process that would eventually end racial segregation in the U.S. Army. The official policy was for African-American soldiers to be gradually integrated into predominately Caucasian units. An exception to that rule, and one of the first units to be integrated at the Presidio, was the Letterman General Hospital Custodial Services Branch, seen in this September 1949 unit photograph. The commander, a Medical Corps officer, sits center front with the unit's first sergeant. The service and combat stripes on the sleeves of many of the men, particularly the noncommissioned officers sitting in the front row, show that these soldiers had many years of Army experience, with some having served in combat.

The famous Presidio Golf Club was established as a nine-hole course in 1895. Over the years, the golf links served as a drill field (see pages 38–39) and 1906 earthquake refugee camp. In the early years, club membership dues were $10 a month, green fees 50¢ a day, and a caddie fee of 25¢ for nine holes. Later expanded to 18 holes, the club had both military and civilian members, including baseball star Joe DiMaggio. Patients from Letterman Hospital could use the course as part of their recovery therapy. Homes on Washington Boulevard face the greens.

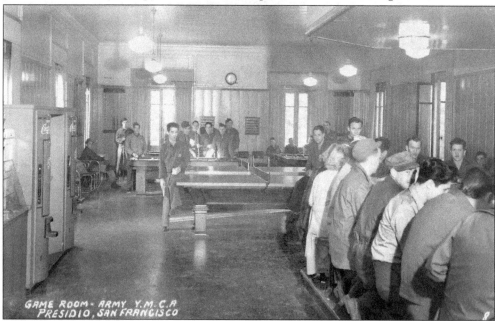

Off-duty troops gather in the game room at the Presidio YMCA (see page 87) in this shot. The "Y" was located across the street from Letterman Hospital, close enough to be used by the recovering patients seen in this photograph at the ping-pong table. In 1954, the YMCA departed the Presidio and the Red Cross took over the building. Originally the Servicemen's Club at the 1915 exposition, the building was torn down in the late 1960s to make room for a parking lot.

118

Major Powers, the Presidio chaplain, and proud parents Master Sgt. Robert Bowen Sr. and his wife, Lucy, pose in front of the Chapel of Our Lady with daughter Linda, newly arrived at Letterman Hospital in December 1949. Two years later, the New England–style chapel would be enlarged and extensively remodeled.

Santa Claus brought gifts but no reindeer to the children's annual Christmas party at the Presidio's Enlisted Service Club (Golden Gate Club). The club, built in 1949, was the scene of a number of significant events, including the signing of a joint security treaty between the United States and Japan. In this 1952 photograph, many parents can be seen in uniform wearing the six-pointed star of the Sixth Army on their left sleeves. (Courtesy of National Park Service, Golden Gate Recreation Area, Park Archives and Records Center, GOGA-2979.033, William K. Toy Collection.)

The Doyle Drive (Highway 101) approach to the Golden Gate Bridge cuts through the Presidio in this 1950s aerial view. World War II barracks and a parking lot fill the lower portion of the photograph over what was once the west end of Crissy Field (see page 90). The newly built Sgt. Roy Harmon U.S. Army Reserve Center is to the left of Stilwell Hall. The pet cemetery is underneath the freeway behind Stilwell Hall. Behind Doyle Drive are the red brick stables, and on the hill above the stables is the Religious Activity Center, originally a cavalry barracks. The buildings of Fort Winfield Scott can be seen in the upper right corner. (Courtesy of the National Park Service, Golden Gate Recreation Area, Park Archives and Record Center, GOGA-2266.)

The Sixth Army Band and Honor Guard participate in ceremonies along the railroad track for a blood drive. The special military blood procurement car sits on the siding near the parked aircraft of the Sixth Army Flight Detachment. The planes and helicopter were used for liaison and MedEvac flights. In 1974, Crissy Field was closed to fixed-wing aircraft. The San Francisco Belt Railroad delivered its last freight cars to the Presidio in 1972. The last time the train track was used was for the Bicentennial Train exhibit in 1976.

During the height of the Vietnam War, the old Letterman Hospital was replaced by a new modern facility. The 550-bed hospital was designed by San Francisco architects Milton Pflueger and Douglas Stone. A solarium with bay views was the favorite room on each floor. In 1970, the hospital cared for over 900 in-house patients a month. The building was demolished in 2003 to make room for the new Letterman Digital Arts Center. (Courtesy of Will Elder.)

During the Vietnam War years, the Presidio experienced many demonstrations at its gates. In this July 1968 photograph, demonstrators in support of the Nine for Peace gather at the Lombard Gate with their portable statue of Saint Francis of Assisi. The "Nine for Peace" were war resisters from all four branches of the military who had sought sanctuary in a San Francisco church. Some of the nine were held in the post stockade at Fort Winfield Scott (see page 73).

The Presidio Mutiny began in October 1968 at the Fort Scott stockade. The stockade was overcrowded with all sorts of military prisoners. Deserters, AWOLs, war resisters, barracks thieves, and drug users were all part of the mix. After a prisoner was shot and killed trying to escape, 27 prisoners refused to work and staged a sit-down strike. Anti-war activists enthusiastically supported the prisoners. These two celluloid protest buttons promote an April 6, 1969, march on the Presidio and a call to free the "Presidio 27."

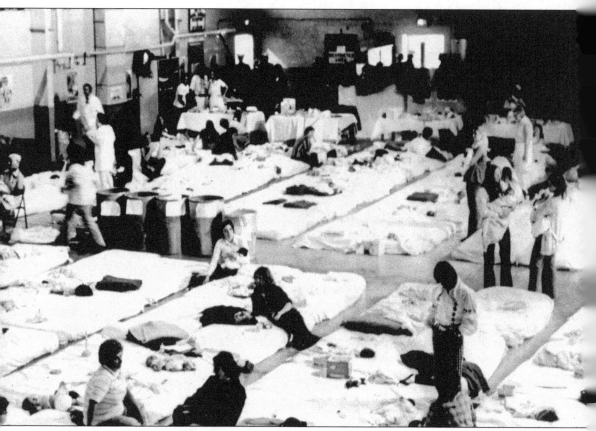

The war in Vietnam ended in April 1975 with the fall of Saigon to the North Vietnamese Army. That month, the Presidio became the destination for over 1,300 tiny refugees from Vietnam. Officially known as Operation SPOVO (Support of Vietnamese Orphans), but popularly known as "Operation Baby Lift," it airlifted infants and toddlers by World Airways to California. At the request of the American Voluntary Agencies arranging for the adoptions of the Vietnamese children, the Sixth U.S. Army arranged to house and feed the children at the Harmon Hall Reserve Center (page 120) until permanent accommodations could be provided for each child. Army and civilian volunteers turned the large drill hall into a nursery. Mattresses were spread out on the floor so the children wouldn't roll off, and sheets, pillows, blankets, as well as baby and medical supplies, were provided. Units stationed at the Presidio responded, providing doctors, nurses, cooks, drivers, and security. Twenty-four hour shifts were not uncommon. This photograph by Jim Stuhler, of the *Star Presidian* newspaper, shows the infants and volunteers in the large gymnasium-like facility. The last babies to be processed departed the Presidio in May. (Courtesy of the National Park Service, Golden Gate Recreation Area, Park Archives and Records Center, GOGA-1766.)

In a major reorganization of the U.S. Army in 1973, the Sixth Army relinquished the command and control of active-duty forces within its area of operation. The Sixth's new mission was for the supervision and readiness of Army Reserve and National Guard units training within the western United States. In this photograph, Army Reserve First Sgt. Gabriel Harp of the Presidio's 353rd Psychological Operations Battalion takes an opportunity to shave while on a field training exercise (FTX).

Army reservists played a greater military role in the years following the end of the draft and the establishment of an all-volunteer Army. Many reservists had previously served active-duty enlistments. Staff Sgt. Santok Singh Sandhu of the 353rd Psychological Operations Battalion was a veteran of the Army's 1st Infantry Division.

Presidio of San Francisco

United States
1847 - 1994

Spain
1776 - 1822

Mexico
1822 - 1847

**Inactivation/Retreat Ceremony
September 30, 1994**

In 1988, the Base Realignment and Closure Commission recommended that the Presidio be closed as an active military installation. On September 30, 1994, the Sixth Army and Presidio Garrison held an Inactivation/Retreat Ceremony, bringing to an end the Presidio's 218-year history as a military post. Illustrated on the event program were the coat of arms of the Kingdom of Castile and Leon (long used as the official crest of Presidio) and the flags of Spain, Mexico, and the United States.

While the active Army has moved on, the souls buried at the San Francisco National Cemetery

The era when the Presidio was "the best duty station, bar none, in the United States Army" was over. The 1972 law that created the Golden Gate National Recreation Area required the Presidio's conversion to National Park status once the Army post was closed. On October 1, 1994, the official ceremony was held transferring the Presidio to the National Park Service, beginning a new era as the only urban national park in the United States. The Presidio of San Francisco would continue to be a special place, bar none.

THE PRESIDIO
FROM POST TO PARK

remain as silent sentries. . . .

BIBLIOGRAPHY

Baylis, Capt. C.D., and First Lt. Miller Ryan. *Historical and Pictorial Review of the Harbor Defenses of San Francisco*. Baton Rouge, LA: The Army and Navy Publishing Company, Inc., 1941.

Bearss, Edwin C. *Fort Point, Historic Structure Report, Historic Data Section*. Denver: National Park Service, 1973.

Benton, Lisa M. *The Presidio, From Army Post to National Park*. York, PA: Northwestern University Press, 1998.

Brack, Mark L., James P. Delgado, and Waverly Lowell. Presidio of San Francisco National Historic Landmark District, *Historic American Buildings Survey Report*. San Francisco: National Park Service, 1985.

Chin, Brian B. *Artillery at the Golden Gate, The Harbor Defenses of San Francisco in World War II*. Missoula, MT: Pictorial Histories, 1994.

Delahanty, Randolph. *San Francisco, the Ultimate Guide*. San Francisco: Chronicle Books, 1989.

Givens, James D. and John A. Martini. *The Presidio's Photographer*. Fairfax, VA: National Park Service Monograph GGNRA Archives, 2000.

Haller, Stephen A. *The Last Word in Airfields, San Francisco's Crissy Field*. San Francisco: Golden Gate National Parks Association, 2001.

Haller, Stephen A. *The Last Word in Airfields, A Special History Study of Crissy Field, Presidio of San Francisco*. San Francisco: National Park Service, 1994.

Hirasuna, Delphine. *Presidio Gateways: Views of a National Landmark at San Francisco's Golden Gate*. San Francisco: Chronicle Books, 1994.

Langellier, John Phillip and Daniel Bernard Rosen. *El Presidio de San Francisco, A History under Spain and Mexico, 1776–1846*. Denver: National Park Service, 1992.

Listening Post. *The History of Letterman General Hospital*. Presidio of San Francisco, CA 1919.

McKane, John and Anthony Perles. *Inside Muni, The Properties and Operations of the Municipal Railway of San Francisco*. Glendale, CA: Interurban Press, 1982.

Presidio of San Francisco, 1776–1961, Directory and Guide. Lubbock, TX: C.F. Boone Nationwide Publications, 1961.

Shamburger, Page and Joe Christy. *Command the Horizon, A Pictorial History of Aviation*. New York: Castle Books, 1968.

Sullivan, Charles J. *Army Posts and Towns, The Baedeker of the Army*. Burlington, VT: Free Press Interstate Printing Corporation, 1935.

The Pacific War and Peace: Americans of Japanese Ancestry in Military Intelligence Service, 1941 to 1952. San Francisco: Military Intelligence Service Association of Northern California and the National Japanese American Historical Society, 1991.

Thompson, Erwin N. *Defender of the Gate, Presidio of San Francisco, A History from 1846 to 1995*. Historic Resource Study, 2 volumes. Golden Gate National Recreation Area, California: National Park Service, 1997.

Takaki, Ronald. *Strangers From a Different Shore, A History of Asian Americans*. Boston: Little Brown and Company, 1989.

Todd, Frank Morton. *The Story of the Exposition: Being the Official History of the International Celebration Held at San Francisco in 1915 to Commemorate the Discovery of the Pacific Ocean and the Construction of the Panama Canal*. 5 volumes. New York: G.P. Putnam's Sons, 1921.

Twelfth U.S. Infantry, 1789–1919, Its Story by Its Men. New York: Knickerbocker Press, 1919.

Weigley, Russell F. *History of the United States Army*. New York: Macmillan Publishing Co., Inc. 1967.

Visit us at
arcadiapublishing.com

CPSIA information can be obtained
at www.ICGtesting.com
Printed in the USA
BVHW010141170619
551175BV00011B/72/P

9 781531 616076